STATISTICS MAD
Do It Yourself

K.V.S. SARMA
Professor and Head
Department of Statistics
Sri Venkateswara University
Tirupati

Prentice-Hall of India Private Limited
New Delhi - 110 001
2006

Rs. 175.00

STATISTICS MADE SIMPLE—Do it Yourself on PC
by K.V.S. Sarma

© 2001 by Prentice-Hall of India Private Limited, New Delhi. All rights reserved. No part of this book may be reproduced in any form, by mimeograph or any other means, without permission in writing from the publisher.

ISBN-81-203-1741-6

The export rights of this book are vested solely with the publisher.

Fourth Printing ... ... **June, 2006**

Published by Asoke K. Ghosh, Prentice-Hall of India Private Limited, M-97, Connaught Circus, New Delhi-110001 and Printed by Jay Print Pack Private Limited, New Delhi-110015.

STATISTICS MADE SIMPLE
Do It Yourself on PC

*To
my mother—
the real mother board in me*

Contents

Preface

1. **THE ROLE OF STATISTICS IN RESEARCH**   1–27
    1.1   Statistics in Research   *1*
    1.2   Research in Statistics   *2*
    1.3   Common Statistical Issues in Research   *3*
    1.4   Data Collection   *4*
        1.4.1   Survey Method   *5*
        1.4.2   Experimental Method   *7*
    1.5   Coding of Data   *8*
    1.6   Tabulation and Presentation of Data   *9*
    1.7   Some Case Studies in Statistical Analysis   *10*
    1.8   The Statistics Toolkit 1   *11*
        1.8.1   Tabulation of Data—One-way Frequencies   *11*
        1.8.2   Cross Tabulations—Two-way Frequencies   *13*
        1.8.3   Histogram—The Graphic Way of Describing a Variable   *15*
        1.8.4   Bars, Lines and Pie Diagrams as Visual Aids   *16*
    1.9   The Statistics Toolkit 2   *19*
        1.9.1   Summary Statistics   *19*
        1.9.2   The Population and the Sample   *23*
        1.9.3   Sample Statistics and Their Formulae   *24*
    1.10   Role of Computer in Statistical Analysis   *26*
    References   *27*
    Suggested Readings   *27*
    Do It Yourself   *27*

2. **BASICS OF A COMPUTER**   28–41
    2.1   The Evolution of Computer   *28*
    2.2   The Personal Computer (PC) and Its Components   *29*
    2.3   Peripheral Devices for the Computer   *32*
    2.4   Getting Started with the PC   *36*
    2.5   Creating a Directory (Folder)   *36*
    2.6   Getting on to FoxPro   *38*
    2.7   DOS or Windows?   *39*
    References   *40*
    Do It Yourself   *40*

CONTENTS

3. **DATA HANDLING AND STATISTICS THROUGH FOXPRO**   42–64
   - 3.1 How to Create a Data File? *42*
   - 3.2 Data File for the Tribal Food Problem *44*
     - 3.2.1 Changing the Structure *46*
     - 3.2.2 Editing the Data File *48*
     - 3.2.3 Printing the Data File *50*
     - 3.2.4 Statistical Calculations with FoxPro *51*
     - 3.2.5 Frequency Tables and Cross-tabulation *53*
   - 3.3 FoxPro File for Blood-bank Data — A Case of Hospital Management *54*
   - 3.4 FoxPro File for Food Grains Data—A Case in Economics *59*
   - 3.5 FoxPro File for Plant Growth Problem—A Case of Experimental Data *61*
   - *References 63*
   - *Do It Yourself 64*

4. **WINDOWS AND MS-OFFICE FOR RESEARCH STUDIES**   65–83
   - 4.1 About Windows 95/98 *65*
   - 4.2 The Windows Explorer *68*
   - 4.3 Opening FoxPro in Windows *73*
   - 4.4 About MS-Office *74*
   - 4.5 About MS-Word *76*
   - *References 82*
   - *Do It Yourself 83*

5. **DATA HANDLING IN EXCEL**   84–101
   - 5.1 Getting Started with Excel *84*
   - 5.2 The Excel Worksheet *85*
   - 5.3 Data Entry on the Worksheet *86*
   - 5.4 Data Sheet for the Food Grains Problem *88*
   - 5.5 Calculations on the Worksheet *90*
   - 5.6 Built-in Functions for Quick Use *92*
   - 5.7 Working with the Tribal Food Problem in Excel *94*
   - 5.8 Operations on Tables *96*
   - 5.9 Printing the Data and Results *98*
   - *Reference 99*
   - *Do It Yourself 99*

6. **GRAPHS AND CHARTS IN EXCEL**   102–114
   - 6.1 Construction of a Bar Chart *103*
   - 6.2 Chart Options *104*
   - 6.3 Construction of a Pie Chart *107*
   - 6.4 Construction of a Line Chart *108*

6.5 Cut, Copy, Paste and Print Options for Graphs *111*
References 113
Do It Yourself 113

7. **DESCRIPTIVE STATISTICS USING EXCEL** 115–130
   7.1 Data Analysis Park in Excel *115*
   7.2 Descriptive Statistics *116*
      7.2.1 Summary Statistics *116*
      7.2.2 Frequency Distribution and Histogram *119*
   7.3 Cross-tabulations and Pivot Tables *123*
   7.4 The Concept of Probability *127*
   References 128
   Do It Yourself 128

8. **INFERENTIAL STATISTICS USING EXCEL** 131–162
   8.1 Estimation of Unknown Parameter Values *131*
   8.2 Testing of Hypotheses *132*
   8.3 Statistical Tests Concerning Means *136*
      8.3.1 The One-sample Z-test for Mean *136*
      8.3.2 The One-sample t-Test for Mean *137*
      8.3.3 The Two-sample Z-Test for Means *139*
      8.3.4 The Two-sample t-Test for Means *141*
      8.3.5 The Paired t-Test *143*
   8.4 The F-Test for Variance *145*
   8.5 Analysis of Variance *146*
      8.5.1 One-way ANOVA *147*
      8.5.2 Two-way ANOVA with Replication *151*
      8.5.3 Two-way ANOVA without Replication *154*
   8.6 The Chi-square Test *155*
      8.6.1 Chi-square Test for Goodness of Fit *155*
      8.6.2 Chi-square Test for Independence *157*
   References 159
   Do It Yourself 159

9. **CORRELATION AND REGRESSION ANALYSIS** 163–183
   9.1 Correlation Analysis *163*
   9.2 Simple Regression Analysis *167*
   9.3 Multiple Linear Regression *174*
   9.4 Diagnostic Analysis of Regression *178*
   References 181
   Do It Yourself 181

10. **FOXPRO PROGRAMS FOR QUICK STATISTICS** 184–205
    10.1 Program 1: Creating a Blank Database File *184*
    10.2 Program 2: Descriptive Statistics (DSTAT) *185*
    10.3 Program 3: Creating a Frequency Table (FTAB) *190*

10.4 Program 4: Cross-tabulations (CROSSTABS) *194*
10.5 Program 5: Two Sample Tests for Means *197*
10.6 Program 6: Computation of Correlation Coefficient *200*
10.7 Program 7: Chi-square Test for Independence of Attributes *202*

References 205
Do It Yourself 205

11. TREND ANALYSIS AND RELATED TOOLS IN EXCEL 206–229

11.1 Fitting a Trend Line to the Observed Data *206*
11.2 Polynomial Trends *210*
11.3 Logarithmic, Power and Exponential Trends *212*
11.4 Moving Averages *214*
11.5 Exponential Smoothing *217*
11.6 Linear and Compound Growth Values *219*
    11.6.1 The Forecast Function *219*
    11.6.2 The Growth Function *221*
11.7 Financial Functions and Related Tools in Excel *222*
    11.7.1 Simple Financial Functions *222*
    11.7.2 Ranking of Data *223*
    11.7.3 Random Number Generation *224*
    11.7.4 Creation of Statistical Tables *227*

References 228
Do It Yourself 229

11. SOFTWARE FOR HIGHER STATISTICAL ANALYSES 230–240

12.1 SPSS and Its Applications *230*
    12.1.1 Summary Statistics *230*
    12.1.2 Cross Tabulations *232*
12.2 Inferential Tools in SPSS *234*
12.3 Graphs and Other Analysis Tools in SPSS *237*
12.4 Other Useful Softwares *238*

References 238
Do It Yourself 239

Appendices   *241–254*
Index   *255–257*

Preface

While teaching students of statistics I have observed that they tend to spend more time on understanding the mathematical aspects of statistics with less focus on practical applications of statistical techniques. Theorems and the theory-based problems keep them more engrossed as these can be easily reproduced in an examination. The real problem to a statistician arises when a large volume of data is to be analyzed. For instance, a sociologist may approach a statistician with a problem of analysis having around 40 to 50 variables and 300 cases. The statistical analysis may require the computation of means, variances, correlation coefficients, and so on. A statistician can handle this type of problem only if he/she can work on the computer.

Many professionals who work with data analysis prefer to use statistical software packages such as SPSS, SAS, MINITAB, etc. for analysis. It is my observation that these software tools are generally not available with researchers, possibly because some of them are quite expensive and also need formal training in using them.

Statistics Made Simple—Do It Yourself on PC is written with the objective of bridging the gap between the researcher and the computer. Students of research methodology face considerable difficulties in either carrying out statistical analysis or using the computer. The aim of this book is to develop the competence of researchers in the use of basic statistical techniques with the help of commonly available software such as FoxPro and MS-Excel. This approach will also be useful for MBA students and managers wishing to improve their knowledge and skills in the context of business applications of statistical techniques.

Chapter 1 outlines the role of statistics in research. It describes different types of data collection methods, graphs and charts, and various levels of statistics in research along with a motivational session to use statistics in research. Some live studies carried out by the students and researchers of S.V. University are also included in this chapter.

Chapter 2 contains an introductory discussion on the PC for those wishing to gain a working knowledge of the computer system.

Different peripheral devices are explained along with some methods of creating files and directories.

Chapter 3 has a full session on the use of FoxPro for data analysis. All aspects from the stage of creating a data file to the level of doing some statistical calculations are included in this chapter. File handling, creating folders, copying files, printing files using Windows and MS-Office 97 environment are discussed in Chapter 4.

Chapters 5 through 9 contain a detailed explanation of the tools of statistical analysis with the help of MS-Excel. Several training courses on MS-Office deal with the general aspects of Excel and omit its potential usage for solving problems on statistics. These five chapters are meant for covering this gap. Data analysis, graphs, descriptive statistics, histograms, tests of hypothesis, ANOVA, correlation and regression analysis are the major topics covered in these chapters.

In Chapter 10 a set of FoxPro programs for solving simple statistical problems is given. These programs are designed to work with data created in a FoxPro file instead of inputting data from the keyboard. The reader can load these programs into the PC and work with them. The idea is to convey the message that programming knowledge helps in producing custom-designed output, instead of using a predefined form of output provided by standard software.

Chapter 11 describes Excel tools for the study of analysis of trends, forecasts, growth functions, financial functions, etc. Random number generation and its applications are also outlined in this chapter.

Chapter 12 presents an outline of SPSS and its use in advanced statistical analyses. It is treated here briefly, only to impress upon the researcher that certain advanced multivariate analyses can be carried out only with the help of exclusive software like SPSS for statistical applications.

At the end of each chapter, exercise problems are given with the caption 'Do It Yourself'. The aim is to drive the reader in the direction of working on the PC instead of just reading a text.

Although FoxPro is discussed with DOS platform, the file handling aspects and the programs discussed in this book will work on other platforms too, such as Visual FoxPro or FoxPro for Windows. For large-scale data processing, a dot-matrix printer is still used by many institutions and FoxPro for DOS would, therefore, be a right choice as against printing in the Windows environment which requires a laser printer. Similarly, the Excel features discussed would work in Office 2000 version as well.

I am greatly indebted to many of my colleagues in S.V. University who have encouraged me in writing this text. I am particularly grateful to Prof. P. Chengal Reddy of the Department of Anthropology, S.V. University for his constant guidance and sharing of ideas during the preparation of manuscript. I also thankfully acknowledge the help rendered by Prof. E. Satyanarayana, Head, Department of Statistics, who whole-heartedly inspired me to work on this project and allowed me to use the departmental facilities.

This book would not have taken this shape without the moral support of my wife Sarada and my sons Karteek and Srikanth. I am grateful to them for their understanding and forebearance during the long hours I spent working on the manuscript.

Despite the help I received from several quarters including that from my publishers, I myself hold the responsibility for any shortcomings in this book. Suggestions for improvement of the book from teachers and students alike will be gratefully received.

K.V.S. Sarma

The Role of Statistics in Research

Some people hate the very name 'Statistics', but I find them full of beauty and interest. Whenever they are not brutalized, but delicately handled by the higher methods, and are warily interpreted, their power of dealing with complicated phenomena is extraordinary.
— *Sir Francis Galton*

1.1 STATISTICS IN RESEARCH

Statistical analysis is a vital component in every aspect of contemporary research. Social surveys, laboratory experiments, clinical trials, etc., require statistical treatment of the findings before arriving at valid conclusions. The research requirements today call for the use of statistical techniques in every field of knowledge. The findings of any research have to be justified in the light of statistical logic.

Statistics is studied as a component of regular curriculum in several courses at both undergraduate and postgraduate levels. A course on research methodology for doctoral programme includes sampling design, data collection, statistical analysis and report writing procedures.

It is generally observed that researchers collect data according to their plan of research, objectives and limitations and then try to find out which statistical tools would suit the data and give a rich presentation. Though this approach is not completely wrong, it stands most often, as a 'technique orientation' rather than 'objective orientation' and gives an impression that statistical tools have been used just because they are available. It is essential to take into account a statistical design before collecting data, especially with respect to sampling.

It is not necessary to apply advanced statistical tools in every data analysis; certain tools may not be applicable in some cases. Simple statistics like averages, percentages and standard deviation would reveal a great information in many observational studies. Exploratory investigations may, however, require some advanced tools.

With the availability of standard computer software, it is now an easy job to 'compute' all the statistical parameters with simple office software, though some special softwares are available exclusively for statistical work. The researcher has only to decide which tool to be used for a particular analysis and leave the rest of the job to the computer. Graphical aids from computer would also help in proper interpretation of the results.

Several researchers treat 'statistics' as a tool only to represent their research findings in the form of tables and graphs. This is a conventional approach and commonly found with administrators to display numerical facts of their organization. But statistics has something more than this. It is an inferential science. It is a science of decision-making which helps to find out the *truth* from the available figures (data). It is the only way out to take decisions in the face of uncertainty.

1.2 RESEARCH IN STATISTICS

A lot of research in statistics has taken place in the 20th century and new research is still going on in different areas of application. Research in statistics has two dimensions: theoretical and applied. All the statistical methods used by a researcher have a strong mathematical basis. Using the foundations of probability theory, statistics has evolved as 'a science in search of truth'.

The problems posed by researchers in various branches of science like biology, agricultural research, clinical trials, anthropological studies as well as problems in economics leading to planning, forecasting, etc., have made theoretical statisticians work on suitable methods to handle quantitative data. This has given rise to a large number of analytical tools, which are popularly used in applied research. In a way, many theoretical results have also come out of a practical need to answer a researcher's questions.

The contributions made by Indians in the field of statistics are worth noting at this stage. While Ronald A. Fisher is treated as the *Father of Statistics* by the community of statisticians all over the world, the Indian scenario has the privilege of respecting Prof. Prasant Chandra Mahalanobis as the father of statistical science in India. He has founded the Indian Statistical Institute (ISI) in 1931 and created an opportunity to a number of scientists to work in the field of both theoretical and applied statistics. The vision of Mahalanobis was so illuminating that the work done by the Indian statistical community could find a place not far away from the centre of the world's statistical map in the 20th century. While the important contributions made by Fisher were in the direction of scientific research and experimental methods, Mahalanobis viewed statistics as a tool in increasing the efficiency of all human efforts and also concentrated on sample surveys [Rao, C.R. (1993)]. Mahalanobis is known for his famous work on an important statistic known as D^2 statistic, which is very popular among social scientists. During the tenure of Mahalanobis, several statisticians from abroad including R.A. Fisher, A.N. Kolmogorov, J. Neyman, W.A. Shewart, W.E. Deming and Abraham Wald have visited ISI, shared their thinking with Indian statisticians and delivered lectures on different topics.

According to Mahalanobis, "statistics must have a clearly defined purpose, one aspect of which is scientific advancement and the other, human welfare and national development". Every researcher and user of statistics should notice this fact to avoid misuse of statistics.

Prof. C.R. Rao is another pioneering Indian statistician who made significant contributions to the theory and application of statistics. His contributions are well known in the field of statistical inference and multivariate analysis. At ISI, he was instrumental for the development of several analytical tools and for the development of computer-oriented methods in statistical analysis. A prominent Indian journal on statistics, *Sankhya* is edited by Rao and considered as a powerful medium for statistical communications.

A statistician is therefore like a server, examining, understanding, participating and serving all the statistical needs of researchers from different fields. However, the quantum of statistics in contemporary research still seems to be lagging behind. One reason for this appears to be the failure of the researcher in studying the problem from a quantitative angle. The gap between theory and application is still wide and should be reduced. The most powerful medium to bridge this gap is a 'computer software'. It is therefore high time for researchers to understand the computer-oriented aspects of handling statistical data and to perform routine investigations on their own.

The objective of this book is to explain some tips for statistical analysis and to highlight how an ordinary office computer can be used to carry out basic statistical analysis for research work.

1.3 COMMON STATISTICAL ISSUES IN RESEARCH

There are different types of statistical issues faced by a researcher. One may broadly classify them into the following groups according to the stage of research:

Level 1: Data collection and recording stage
- Sampling scheme of a survey
- Layout of an experiment
- Data coding, scoring and recording
- Tabulation and presentation of data.

Level 2: Computing basic statistics
- Proportions and percentages
- Average and standard deviation of variables
- Measures of consistency of data
- Frequency distributions and histograms
- Measures of location (averages), variation and shape
- Cross tabulations.

Level 3: Statistical tests of hypotheses
- Comparison of means of independent groups
- Comparison of means of paired values
- Comparison of proportions
- Comparison of variances.

Level 4: Associations and relationships
- Tests of independence between attributes (count data)
- Contingency and association measures
- Correlation and regression
- Non-parametric methods.

Level 5: Multivariate methods
- Factor analysis
- Cluster analysis
- Discriminant analysis
- Probit and logit analysis
- Path analysis
- Profile analysis
- Multivariate ANOVA
- Analysis of factorial experiments.

Each of the above aspects and tools requires a fundamental understanding of its statistical origin and purpose. In the following section, we examine some aspects of data collection for a research study.

1.4 DATA COLLECTION

Data for research is usually obtained by two methods. One is by conducting a survey of the entire population in which the researcher is interested or by taking a sample of it. The other one is by conducting laboratory or field experiments and generating data. Each of these methods requires some statistical knowledge.

1.4.1 Survey Method

The following issues have to be resolved when the data for a research problem have to be obtained from a survey.

What is a sample?

According to Snedecor and Cochran (1994), "a *sample* is a set of items or individuals selected from a larger aggregate or population about which we wish quantitative information". Sampling is the process of drawing samples from a given population. The results obtained from a sample will be of interest only if they convey something about the population including that portion which we have not studied in the sample.

Sample or census?

If all the individuals or units of a population are inspected for the study, it is called *census* or *100% inspection*. It is usually difficult to carryout a true census that involves a lot of money, manpower and time. Further 100% inspection is sometimes impossible and may lead to nonsense like the following message.

> In order to know the blood glucose level one should pump out the entire blood into a vessel and get it tested in laboratory. This is what 100% inspection calls for.

No one would agree for this proposal. So, the only way is to take a sample. In the context of 'quality control', census is called 'screening inspection', according to which, every item is inspected with respect to some 'vital' characteristics.

A census becomes essential in some cases where the units to be inspected are a few and the nature of the problem requires data from each such unit. Here is a situation of this type.

> There are only 10 sugar factories in a region. If the study requires some vital information about this industry, it is necessary to include all of them.

So, census is the only way.

What type of sample should be taken?

The first point to resolve is whether the sample should be a *random* sample or a *purposive* sample. A sample, in which each unit of the population has an equal chance of being selected into the sample, is called a 'random sample'. Such a sample will be free of investigator's bias and the inferences based on such samples will be unbiased. A 'purposive sample' is a non-random sample which is based on factors like convenience in data collection, budget and time constraints and so on. 'Opinion surveys' conducted by popular magazines among their readers are examples of such samples. Though it is likely to be influenced by the investigator while selecting the units into the sample, still some practical applications require a judgement sampling.

Many statistical tools like tests of significance cannot be applied when the sample is not a random one. The data from a non-random sample can be presented for understanding the situation, but inferences about the unknown portion of the population cannot be drawn from such samples.

How should the sample be? What should be the size?

One should remember that the sample should be a true representative of the population. A poor representative naturally gives a poor show! All the results based on them lead to wrong inferences. The size of the sample to be taken is determined by factors like the nature of the population, the purpose of the study, the desired accuracy in the results and the budget available. Practitioners believe in thumb rules like taking 10% of the population as the sample. Statistical methods based on probability theory help in the proper determination of the sample size.

Sampling unit

The basic unit from which the researcher wishes to extract data shall be carefully determined. It could be an individual like a student from a school, a patient from a hospital or a taxpayer in a locality. Such a unit is known as the sampling unit. The unit could also be a team of respondents headed by an individual, like a family in a village, a department, a school, etc. The sampling unit may be different if the sample is taken in different stages. For instance, in the first stage, the district may be the unit, in the second stage a *mandal* (group of villages) could be the unit and within *mandal* a village and latter a household could be the unit for sampling. Beri (1989) has discussed some interesting aspects of sampling in the context of marketing research.

The nature of the population and the method of sampling

The population from which the sample has to be taken shall be examined carefully. If all the units were homogeneous, it would be enough to collect data through a simple random sample. In order to take a random sample it is a good practice to follow the steps given below:

(i) List out the population units and identify them as 1, 2, 3, ..., N
(ii) Refer to the table of random numbers from any statistical tables
(iii) Read out the numbers in some order and choose the population units in that sequence, avoiding repeated numbers
(iv) Stop, when the required number of units get selected.

This method is useful only if the population units are distinct and listed in some order. The list of villages in a district is one such example. Suppose we have a box of electrical switches having 1000 pieces and we are asked to take a 10% random sample out of it. This may require a complete unpacking of the lot. Special efforts are made while packing such items so that from the main lot, a sample of sub-lots (say, of 10 each) can be first selected at random and from them the pieces can be inspected.

If the population is not homogeneous with respect to one or more characteristics, then a method called *stratified random sampling* is recommended. It takes care of the groups (called *strata*) within the population and suggests a method of drawing the sample in such a way that the different groups will have proper representation in the sample. The stratification can be made on the basis of factors like caste, residential status like rural or urban, gender, etc.

Primary and secondary data sources

The data obtained directly from each unit in a survey is often known as *primary* data. Collection of primary data from the respondents requires a lot of skill. In this method, the investigator would interact directly with the respondent and obtains the information.

In certain research problems, data from external sources will be required. Some government and private agencies are involved in maintaining large databases. The user can access them through published material or directly from computer media.

1.4.2 Experimental Method

Laboratory experiments, clinical trials and field research are conducted under controlled conditions. As such there will not be a population of the desired type from which a sample could be taken. The experiment is a source for generating data. The interesting aspect of experimental data is that usually a small number of observations would be available unlike in the case of sample surveys. Using these few values, one has to draw valid conclusions not only for the experimental environment but also for the entire hypothetical population, which satisfies the experimental conditions.

The concept of 'experimental design' is something essential for any researcher before going ahead with the investigations. A design is basically a *plan* or *scheme* according to which, data should be collected from the experiment. It specifies the number of trials required, the way in which the treatments (factors) should be allocated to various experimental subjects and how the subjects should be organized into homogeneous groups. All these are aimed at reducing the researcher's bias and to estimate the effect of treatments and their interactions in an unbiased manner.

In many clinical and laboratory experiments, the investigator applies a treatment like a new drug at different doses and wishes to compare the response with something *standard* or *control*. The data for such experiments should be recorded in a systematic way. A lot of research has taken place in the field of design of experiments. The pioneering effort was by R.A. Fisher (1890–1962) whose early studies have taken place in the field of biometrics and agricultural experiments. These experimental designs have been found very useful in the field of statistical quality control (SQC).

In any study based on either sampling or experiments, the investigator has to record the data in such a way that it is amicable for quantitative analysis.

1.5 CODING OF DATA

Statistical data can be broadly classified into two categories. The first one is *qualitative* data, which signifies an attribute like affection, gender, level of satisfaction, colour, taste, etc., which cannot be expressed on an interval scale or in some measured units. The second one is *quantitative* data, which signifies a measurement on an interval scale like body mass index, blood cholesterol level, plant height, distance, income of a person, etc.

Any aspect on which data is collected is called a *variable*. Usually all the independent questions in a survey are attached to separate variables. It is necessary to determine whether a variable should be taken as a *count* or *measurement*.

Coding is a method of assigning a numeric value or a symbol to each level of the attribute under consideration. All the data for statistical analysis are required in quantitative form only. Even if the data is qualitative, it has to be 'coded' before carrying out statistical analysis and latter 'decoded' for interpretation.

For a qualitative variable, it is sometimes convenient to record data in non-numeric form like 'M' for Male and 'F' for Female. One can also put a '✓' for Yes and '✘' for No. When it comes to analysis, these symbols or letters do not work. It is therefore necessary to assign a *numeric code* to them. There can also be more than two levels of response and they require a multilevel code. For instance, the level of agreement over an issue may be coded like giving '–2' for Strongly Disagree, '–1' for Disagree, '0' for Indifferent, '1' for Agree and '2' for Strongly Agree.

For quantitative variables, the data can be recorded in the same way it was collected. However, it would prove useful to classify the actual values into convenient groups and carry out the analysis on the groups itself. For instance, one may use the code '1' if the monthly income is less than 1000, code '2' if it is between 1001 and 2000, code '3' if it is between 2001 and 3000, and so on.

Now the data will be ready for primary analysis like preparation of tables and graphic representation.

1.6 TABULATION AND PRESENTATION OF DATA

Tabulation is basically a job of summarising the raw data into an understandable form. It is difficult to say which is raw data and which is a finished one. The data collected from the field and recorded in the way it was collected would be *raw data*. After tabulation we may find a level of clarity or a sort of summary of what has been collected.

The tabulated data of one researcher may serve as raw data for a different researcher. Monthly reports of trade and commerce, food stocks and hospital statistics are some examples. These tabulated data may need further 'finishing' before interpreting them. One should note that a tabulated data is like a 'semi-finished product' and lacks several primary details!

Tabulation can also be defined as a process of counting the number of cases falling in different categories of a variable. Hand calculations would help only in case of small sized data to tabulate with a few variables but computer tabulation is the most convenient way of doing this job. Once the data is fed to the computer, several basic queries can be solved without any difficulty.

The next job of the researcher is to determine the most effective way of presenting the data. *Tables* and *graphs* are the first choice. With the present facilities on computer, it is possible to present statistical data very effectively. Diagrammatic presentation using different icons is a 'catchy' way of presentation. *Vertical bars, lines* and *pie diagrams* are the primary choice of any researcher. Some special diagrams like *stem-and-leaf* diagram, *scatter* diagram and *3-D surfaces*, are often used to present research findings in a lucid way.

Until the emergence of a personal computer (PC) and sophisticated software, the aspect of drawing graphs was a difficult and a time-consuming job for researchers. Today, the PC would not only offer a big menu of charts and graphs but it is possible to place them in the text at any place and in any size.

Graphs are essential for another reason. Suppose a researcher wants to find the growth rate of an economic index over a period of time. The first thing is to examine the trends shown by the data. If the data shows a linear trend, a simple linear growth rate should be calculated. If the trend is not linear but closely represents a growth curve, the compound growth rate would be suitable. With the help of the computer it is easy to study the trend on the graph and then a suitable formula can be used to work out the growth rate.

1.7 SOME CASE STUDIES IN STATISTICAL ANALYSIS

The purpose of this section is to describe some cases in which statistical analysis is required to draw conclusions. It is important to note that statistical treatment of the project should be started right from the stage of sampling design and framing of hypotheses. Total dependency on computer for presentation of data is not always correct, though a large number of options are available for a scientific presentation. It is the researcher who has to determine the type of questions to be answered and the statistician would advice him on the type of analysis.

The following are some case studies in which practical issues related to data handling and statistical analysis are discussed. The details of the data and how it should be taken to the computer are discussed for each case in the following chapters.

Case 1: Tribal food habits (a case in nutrition and sociology)

A researcher in nutrition has undertaken a study on the food habits of tribal men and women. The tribal persons are classified according to their group namely *Sugali*, *Yanadi* and *Yerukala*. The data contains tribe information, gender, dietary intake, energy spent on various activities, etc., along with several social factors.

The objective is to study the health status among these tribes in terms of the energy balance and to test whether the difference, if any, can be attributed to factors like tribe, gender. "How to take this data to the computer and what statistical tools are applicable to meet the objectives" is the interest of the researcher.

Case 2: Blood bank performance (a case in hospital management)

This is a case study in which the performance of blood bank in a hospital has to be evaluated. During the period 1993–96, the researcher has collected the data on the number of blood donations received by the bank and the number of transfusions carried out. There are four types of blood groups denoted by O, A, B and AB groups. In each group, there will be positive and negative types depending on the Rh factor. The data shows the monthly number of collections as well as transfusions made.

The problem is to carry out statistical analysis of the performance of the blood bank and draw conclusions.

Case 3: Trends in food grain storage (a case in economic forecasting)

The Food Corporation of India (FCI) is the country's largest stockkeeping unit of food grains. (Over a period of time this author has collected the data on the actual procurement of different food grains.)

It is desired to study the trends in procurement and to forecast the future values in a scientific way. It should be noted that the data is not a random sample of a large population. It is the only available data and should be represented by some basic statistical parameters and graphs.

Case 4: A designed experiment for plant growth (a case in botany)

A botanist has conducted an experiment that contains 4 treatments and each treatment is tried on 2 different cultivars (plant varieties). Under each cultivar, each treatment has been applied on 5 homogeneous subjects (seeds). The growth parameters of the plant like shoot length and plant height, have been recorded at different stages indicating the days after sowing (DAS). The results have been observed at 4 different stages of plant growth.

Now the researcher wants to carry out statistical analysis of the experimental findings. This includes computation of mean, standard error, percentage growth and relationships among variables, tests of hypotheses regarding the effect of factors, etc. Above all, the primary requirement is to take the raw data of the experiment to the computer in a systematic way.

The cases mentioned above are only indicative of the nature and complexity of real world research problems. Proper care is necessary while drawing inferences from the available data and this requires some basic understanding of statistics and the use of computers in handling the data. The computer is the most powerful facility to perform these calculations accurately and in short time. The details of these cases are discussed in latter chapters.

1.8 THE STATISTICS TOOLKIT 1

In this section, we look at some basic aspects of statistical analysis at the primary level. It would be like a *first aid* and provides a basic insight into the type of analysis required at latter stages.

1.8.1 Tabulation of Data—One-way Frequencies

The fundamental requirement in data analysis is that of counting *how many times* each *distinct value* of a variable has occurred. The count or the tally of such values is called the *frequency*. When we arrange the values and their corresponding frequencies as a table, we get what is called a one-dimensional frequency table. This job is called *tabulation* and the variable for tabulation could be either numeric or categorical. Consider the following situation.

EXAMPLE 1.1 The number of surgeries performed by a city hospital during a week has been classified under various heads and the counts are given as a frequency table shown in Table 1.1.

Table 1.1 Count of the Number of Surgeries Performed in the Hospital

Type of surgery	Number of cases
Thoracic	56
Prostate	45
Urologic	61
Abdomen	126
Neuro	41
General	102

Here the type of surgery is a categorical variable indicating different categories of surgeries performed. The frequency is the number of cases performed. The actual raw data would probably contain a book record of the type of surgery and we have to count the number of cases under each category. This is often called categorical or qualitative classification. ■

Here is another example where the variable for tabulation is based on a numeric value.

EXAMPLE 1.2 Computer floppy diskettes are marketed in boxes of 10 diskettes each. The *XYZ* Company has conducted a survey on the complaints received by its customers as a part of its marketing research. If a diskette shows an error in reading or writing on its first use, it is treated as a defective. The distribution of the number of defectives observed in a particular city during March 1998 is shown in Table 1.2.

Table 1.2 Distribution of Defective Floppy Diskettes Per Box

Number of defective diskettes in the box	Number of boxes (frequency)	Per cent (%)	Cumulative per cent (%)
0	240	80	80
1	30	10	90
2	15	5	95
3	9	3	98
4 or more	6	2	100

Here the variable is the number of defective diskettes in the box. Sometimes it would be convenient to express the frequencies in percentage and cumulative percentage. These two columns are also shown in the same table.

The cases having 4 or more defectives in a box is just 2%. The management may also wish to find out what per cent of boxes contain at least one defective. The answer comes from the third column as 20%, because 80% of the boxes are free of defectives. ■

When there are too many data values, it may not look meaningful to identify the distinct values and count their frequency. Instead, we divide the possible data values into distinct groups and then count how many values fall in each group. Such groups are called *classes* or *intervals* and the tabulation method is often known as *continuous classification*. Consider the following situation.

EXAMPLE 1.3 As a part of a research study, a psychologist has studied the degree of anxiety faced by students of Intermediate class studying in private and government institutions. Each student has been questioned on various aspects of anxiety and a score is obtained which is an index of the level of anxiety faced by that individual. In all, there were 100 students in the sample. The frequency distribution of scores has been prepared and given in Table 1.3.

Table 1.3 Frequency Distribution of Anxiety Scores

Anxiety score	No. of students	Anxiety score	No. of students
Below 40	4	52–55	15
40–43	5	55–58	13
43–46	11	58–61	10
46–49	14	61–64	8
49–52	17	64 and above	3

The method of preparing such a table is discussed in Section 1.9.3. ■

The preparation of frequency tables is a cumbersome job with hand-calculations and may sometimes lead to errors which consume a lot of time to set-right. So, a computer-based procedure is a suggested method. In Chapter 5, we discuss a method of deriving this table using a simple computer software.

1.8.2 Cross Tabulations—Two-way Frequencies

Suppose we have two categorical variables (called *attributes*) like sex (male and female), family type (joint and nuclear) and the problem is to count how many of the experimental subjects belong to the different combinations. This is a common problem in social surveys and a proper way of entering data into the computer would make this tabulation very simple. Since there are two variables under consideration, we call this a two-way frequency table or *cross-tabulation*.

While preparing such tables, we select one combination of the pair of categories and count the number of data cases falling in that pair. Then another pair is considered and the frequency is calculated. In statistical analysis, such tables are called *contingency tables*. If there are m categories for one variable and the other has n categories, we get an ($m \times n$) table with m rows and n columns. One attribute will be shown along the rows and the other along the columns.

Some researchers present the percentage of cases along with the frequencies for each cell. A *cell* is the intersection portion of a row and a column. An interesting aspect of cross-tabulation is regarding the calculation of percentage. We have the option of putting the percentage with respect to (i) row total (ii) column total or (iii) grand total. It is the researcher and not the computer to determine which one among these three should be chosen. For some applications, the row total is the basis while the column total or the grand total is the basis for finding the percentage in some other cases. Here is a simple illustration of cross-tabulation.

EXAMPLE 1.4 Statman Software Corporation has evaluated the performance of its employees in statistical training and latter rated them according to their level of success in statistical consultations for its clients. The company has classified its 300 employees as shown in Table 1.4. Here the two categorical variables are the job success and the performance in training.

Table 1.4 Cross-tabulation of Employees according to Job Success and Training Performance

Job success	Training performance			Total
	Below average	Average	Above average	
Poor	12	47	18	77
Average	20	70	47	137
Very good	9	40	37	86
Total	41	157	102	300

The table shows that there are only 9 candidates out of 300 who were very good in job success but below average in training performance. We can present them also in per cent in the same table, if necessary. We can also carry out cross-tabulations with more than two variables. This is usually called *nested classification* and forms an important segment of the primary data analysis. Here is one such situation in which cross-tabulation is with respect to more than two categories.

EXAMPLE 1.5 The Statman Corporation wanted the entire classification of employees not only according to job success and training performance but also according to sex and residential status namely rural and urban. The computer output usually do not appear as we present below. We have to rearrange the output as in Table 1.5.

Table 1.5 Distribution of Employees under Different Categories—A Nested Table

Job success	Residence	Training performance						Total
		Below average		Average		Above average		
		Men	Women	Men	Women	Men	Women	
Poor	Rural	2	2	12	8	4	2	30
	Urban	5	3	16	11	8	4	47
Average	Rural	5	4	18	12	9	8	56
	Urban	6	5	24	16	16	14	81
Very good	Rural	2	1	10	6	7	10	36
	Urban	4	2	14	10	11	9	50
Total		24	17	94	63	55	47	300

We can also use an additional column to show the total number of men and women under each category.

1.8.3 Histogram—The Graphic Way of Describing a Variable

The distribution of total frequency into different groups or classes is known as *frequency distribution*. Quite often it is necessary to describe the frequency distribution of a variable in the form of a simple statistical graph called *histogram*. It is a graphic representation of the frequency distribution by vertical bars against adjacent class intervals. Several statistical tests and other procedures are based on the assumption that the variable under study has a typical shape for its histogram. One of the problems is to determine the class intervals. The basic principle is to first find the difference between the largest and the smallest of the data values. This is called the *range*. We have to divide this range into k classes, where k is usually between

5 and 15 but it is not a hard rule. The idea is to avoid too few or too many classes in the distribution. One formula that is described in many fundamental statistics books is

$$k = 1 + 3.322 \log_{10} N$$

where N is the number of data values.

This is known as *Sturges formula*. Sturges (1926) has suggested this as a method of defining the class interval. The class interval c is then automatically determined by the formula $c = Range/k$ and the value is rounded off to a whole number, if necessary. Daniel (1983) observes that this formula is only a guiding principle and minor modifications can be made in defining the classes. The researcher's knowledge about the data is a prime factor in determining the classes or the class interval. But the basic principle to follow is that the classes should be evenly distributed over the range so that not many classes go without any frequency at all.

The class intervals can be taken either by the *inclusive method* like 10–19, 20–29, 30–39 and so on or by the *exclusive method* as 10–20, 20–30, 30–40 and so on. While the first method avoids ambiguity in allocating values to distinct groups, the second method has good mathematical properties. By observing the convention that a value like 30 should be kept in the class 30–40 and not in 20–30, one can avoid any confusion. When we connect the mid-points of all the bars of the histogram with a smooth curve we get what is called a *frequency curve*. This is treated as a basic tool in understanding the data described by the variable under consideration. The shape of the frequency curve is important and any curve observed from a sample data is usually compared with a standard curve known as the *normal curve*. We shall examine it in details in a latter section. A computer can help easily in getting a histogram and a frequency curve. A typical frequency distribution and the histogram appear as shown in Figure 1.1.

1.8.4 Bars, Lines and Pie Diagrams as Visual Aids

Statistical data can be represented by other methods among which, the *bar diagram* is an important and commonly adopted one. It is used to describe the frequency of cases belonging to different categories of variables. Figure 1.2 is a bar diagram of the data referred in Example 1.1.

A *line diagram* is another way of representing data usually to describe the changes over a period of time. The trends in any time series data is normally represented by a line diagram. We can show one or more variables in a line diagram and observe the trends shown by the data. Examples include the study of economic and

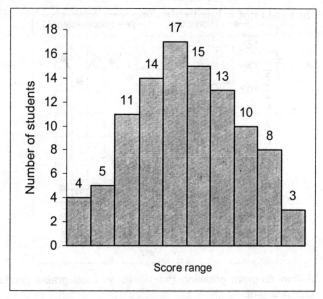

Fig. 1.1 Histogram of anxiety scores.

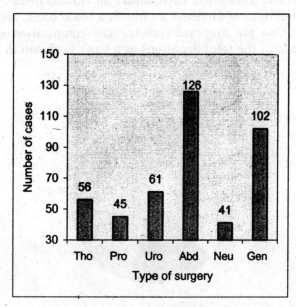

Fig. 1.2 Bar diagram of the number of surgeries performed.

financial data, share market trends, recovery rate of a patient and so on. Figure 1.3 shows a line diagram of trends in the production and procurement of food grains by the Food Corporation of India during 1970–71 to 1980–81.

The *pie diagram* is another popular visual aid to display statistics. It is used when the researcher wants to depict the share of various

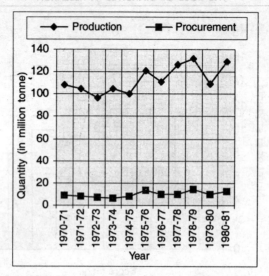

Fig. 1.3 A line diagram showing the trends in food grains production and procurement.

items in a total. Household expenditure on various heads, percentage of blood donations of different groups in a blood bank, etc., are some examples. One pie diagram showing the composition of different blood groups in the total donations at a bank is shown in Figure 1.4.

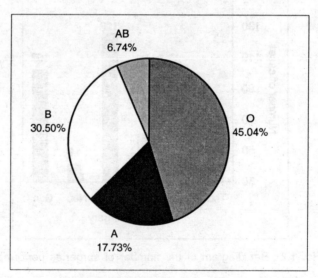

Fig. 1.4 A pie diagram showing the share of various blood groups.

There are several other types of graphs and diagrams, which can be used for a meaningful presentation of statistical data. While there are special computer packages available to handle graphs, MS-Excel is generally sufficient to describe several data patterns.

1.9 THE STATISTICS TOOLKIT 2

Statistical data analysis can be broadly divided into two categories as follows:

Univariate analysis. This aspect deals with *one* variable at a time. Study of descriptive statistics like mean and standard deviation of one single variable at a time, belong to this category.

Multivariate analysis. This aspect deals with *more than one* variable and addresses the problems of inter-relationships among several variables, extraction of hidden factors, problems of classification, etc.

Several statistical tools are available under each category and the researcher has to apply them by proper judgement about the use of the tool. We shall discuss a few of them in the following sections.

1.9.1 Summary Statistics

Descriptive measures of data are very important among the univariate tools. Diagrammatic and graphic representation of data has certain limitations. One can at most understand the implications of a graph but analytical aspects like projections, comparisons and drawing of valid conclusions are possible only when the data is described in terms of quantitative indices. Some commonly used indices are listed in the following section. These are called *summary statistics* and can be classified into three categories as discussed below:

Measures of location

Every statistical data shows a tendency that there will be a *central value* around which the other values are scattered. That value which is at the centre is called a *measure of location* or a *locational value*. There are different measures of location and the commonly used measures are described below:

Mean. It is the arithmetic average of the data values. It is the most widely accepted measure of location and has good mathematical properties.

Median. It is that value of the data below which 50% of the values fall, when the data values are sorted in the increasing or decreasing order.

Mode. It is that value which occurs most frequently in the data. If more than one mode occurs the data is said to be *multi-modal*. In such cases, the *smallest* among the available *modes* is taken as the mode of the distribution. (More details can be seen from the help file of MS-Excel.)

Mean, median and mode are also called *averages*. Each of these measures represents the central tendency of the data. In addition, there are two special averages known as *geometric mean* and *harmonic mean* which have specific applications in biology, demography and in some areas of economics.

For hand calculations, the basic statistical methods suggest the use of classified data instead of raw data. The main reason is that it is difficult to handle large volumes of data with hand calculations. One should remember that by classifying the data into groups and making a frequency table the individual observations loose their identity and cannot contribute to explain the characteristics of the data. As a result, one would get a sort of *interpolated* value and not an *exact* value. But it was the only way when computers and hand calculators were not there. With the advent of computer it is easy to handle any amount of data and the statistics can be computed directly on the raw data.

Measures of variation

In addition to the central value, the *variability* or *dispersion* in the data has to be measured. Variability is basically inherent in every data (unless all values are the same) and indicates the degree of *scatter* or *spread* of values around the central value. Here are a few measures of dispersion.

Range. This is the simplest measure of dispersion and it is the difference between the largest and the smallest values of the data. This is a good measure when the data values are fairly homogeneous. A statement like

> the score of a batsman ranges between 40 and 60.

conveys more information than the statement that

> the score ranges between 0 and 100.

In the latter case, there is an indication that the performance could be unstable.

Variance. It is a measure of the extent to which the actual observations vary from the central value (mean). It is one widely accepted measure of variation and used as a primary tool in data analysis.

Standard deviation (SD). It is the positive square root of the variance. Both the mean and SD are expressed in the same units as those of the original data while *variance* is expressed in squared units. For this reason, practitioners find it convenient to use SD in place of *variance* to explain the dispersion.

Coefficient of variation (CV). It is a relative measure of dispersion and defined as the percentage of the SD to mean. It is a pure number without any unit and hence it can be used to compare the *relative inconsistency* in the data. A lower value of CV indicates relatively more consistent data. Suppose we wish to compare the level of stability in *reading hours* and the *marks* obtained of a student over a series of tests held in 10 months. If the CV for *reading hours* is less than that of *marks*, we say that reading hours of the student are stable than the marks obtained. Similarly if a cricket player scores 60 in one test and 40 in another test, he is better than a player who does 100 in one test and 0 in another test. Note that the average is the same for both of them! In such cases, CV is a very useful analytical tool and helps to assess the stability of data.

There is still another measure of dispersion called the *mean deviation*, which is the *average absolute deviation* of values from the mean. This is however not much used in several conventional applications, but with the advent of computer-oriented methods certain new statistical methods have come up using this concept which is also known as *mean absolute deviation* or MAD.

Measures of shape

While mean and SD are treated as primary tools in data analysis, there are some other properties displayed by a data series. These are related to the shape of the frequency curve with the following measures.

Skewness. It is a measure of the degree of asymmetry of a frequency distribution. If the values below average are more than the values above average, the distribution tends to be *left skewed* and a long left tail appears for the frequency curve. Similarly a *right-skewed* distribution will have a long right tail. For a *symmetric* distribution, the measure of skewness is zero.

Kurtosis. This is a measure of the *peakedness* or *flatness* of the distribution. If the distribution is neither peaked nor flat, then kurtosis is zero. The positive values of kurtosis indicate *too much peakedness* while the negative value indicates *too much flatness* in the shape of the distribution.

The normal distribution

In many real-life situations, the data on the variables are centred at the middle of the distribution. The frequency of values that are different from this central concentration gradually decreases and the frequency curve becomes bell shaped. This is called the *normal curve*. Figures 1.5(a), 1.5(b) and 1.5(c) respectively show a left-skewed, a right-skewed and a normal curve.

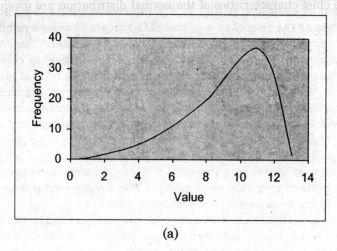

(a)

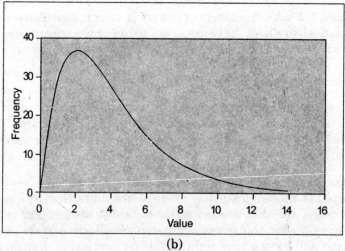

(b)

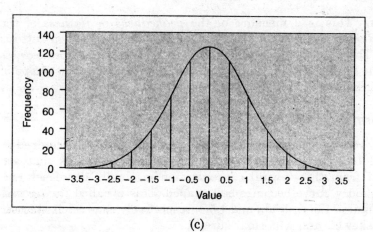

(c)

Fig. 1.5 (a) Left-skewed distribution, (b) right-skewed distribution, (c) normal curve.

The chief characteristics of the normal distribution are given below:

- The distribution is symmetric and neither flat nor peaked
- The mean, median and mode are the same and represent the central value
- About 95% of the data values fall within two standard deviations from the mean
- About 68% of the data values lie within one standard deviation from the mean.

For these reasons, the normal distribution is taken as a reference distribution for describing the nature of an observed distribution. For instance if a sociologist is studying the body mass index (BMI) of individuals of a sample, the frequency distribution of BMI values can be plotted as a histogram and the shape of the frequency curve can be compared with that of the normal curve. If they are close to each other it means that the BMI can be well described by the normal distribution. We may then confidently say that about 5% of the individuals from the selected population will have very extreme values of BMI.

1.9.2 The Population and the Sample

The totality of all cases about which the researcher wants to draw conclusions is called the *population*. All those cases, which are included in the study from the population, will be the *sample*. Statistical methods help in drawing valid inferences about a population using the sample data. The sample taken should be a random one so that the conclusions drawn will be free from bias. Let us look at some technical terms:

Parameters. If the measurements are obtained from the entire population, the values of mean, median, mode, standard deviation, etc., would be *exact* and they are called the *parameters*. Suppose we have a question: "What is the average monthly expenditure of a professor in S.V. University?" If the researcher is able to contact each and every professor and obtain the relevant data, the average will be the *true* value and it is the parameter. If it is not possible, the researcher may take a random sample of few professors and would calculate the average. This need not be the same as the true value. Depending on the method of sampling and the members included in the sample, different possible results could arise, each of them claiming to be an *estimate* of the true value.

Estimates. When the calculated values are based on a sample, the resulting statistics are called the *estimates* of the unknown parameters of the population. The theory of statistics has a special branch called 'statistical inference' that deals with two major issues namely 'estimation' and 'tests of hypothesis'. The theory of probability is the basis for developing methods of estimation based on random samples. A value like the mean obtained from the sample data is often called a *statistic*.

> Unless otherwise specified, the researcher would be always working with sample data and not with census or the entire population data. So, necessary care shall be taken to ensure that the sample is a random sample and the results obtained should also be applicable to the un-inspected portion of the population.

1.9.3 Sample Statistics and Their Formulae

Suppose a raw data contains n values of a variable x. The formulae that can be applied on the data to compute the summary statistics are given below:

$$\text{Mean } \bar{x} = \frac{\sum_{i=1}^{n} x_i}{n}$$

$$\text{Variance } \sigma^2 = \sum_{i=1}^{n} \left[\frac{(x_i - \bar{x})^2}{n-1} \right]$$

$$\text{Standard deviation } S = \sqrt{\frac{n \sum_{i=1}^{n} x_i^2 - \left(\sum_{i=1}^{n} x_i \right)^2}{n(n-1)}}$$

$$\text{Skewness } SK = \frac{n}{(n-1)(n-2)} \sum_{i=1}^{n} \left(\frac{x_i - \bar{x}}{S} \right)^3$$

$$\text{Kurtosis } KS = \frac{n(n-1)}{(n-1)(n-2)(n-3)} \sum_{i=1}^{n} \left(\frac{x_i - \bar{x}}{S} \right)^4 - \frac{3(n-1)^2}{(n-2)(n-3)}$$

The computation of 'median' is based on actual search of raw data after arranging it in ascending or descending order. 'Mode' is also calculated by locating all possible modes and picking up the least value among them. It is worth noting that for a fairly asymmetric distribution there is relationship among the mean, median and the mode given by the formula

$$Mean - Mode = 3(Mean - Median)$$

Similarly one can use several other coefficients for measuring skewness and kurtosis. Since the objective of this book is to drive the reader towards the use of computer, one can directly use the available functions to compute the statistics. More details about this aspect are discussed in Chapter 7.

It should be noted that the above formulae are not applicable for classified data and with the help of computer there is no need to classify the data before finding the summary statistics. If however data is given in a classified form we can write a small program in the computer and workout the statistics.

There is an observation in mathematical statistics that when the data is fairly normally distributed, an unbiased estimate of the population variance can be obtained by using $(n - 1)$ instead of n, in the denominator of the formula. So the estimated variance is given by

$$s^2 = \sum_{i=1}^{n}\left[\frac{(x_i - \bar{x})^2}{n-1}\right] \quad \text{or} \quad \frac{n\sum_{i=1}^{n} x_i^2 - \left(\sum_{i=1}^{n} x_i\right)^2}{n(n-1)}$$

The usual formula for variance is however given by

$$\sigma^2 = \sum_{i=1}^{n}\left[\frac{(x_i - \bar{x})^2}{n-1}\right] \quad \text{or} \quad \frac{\sum x_i^2}{n} - \left(\frac{\sum x_i}{n}\right)^2$$

This estimator s^2 is called the *sample variance*, which is an unbiased estimator of the unknown σ^2. The formulae for the skewness and kurtosis are also based on factors like $(n - 2)$, $(n - 3)$ in the denominator, which represent corrections for small samples. When the sample size is large, the difference between $(n - 1)$ and n in the denominator will be insignificant so that the small-sample formulae can be safely used even for large samples but the other way is not correct.

Standard error. When a parameter like the population mean is estimated from the sample data, it cannot be taken as unique. In fact, if one takes another sample from the same population, a different value of the mean may come out from that data. So, sampling fluctuations can lead to different estimates of the same parameter. If all such possible sample estimates are arranged in the form of a frequency distribution, it is called the *sampling distribution*. The mean of the sampling distribution is in general equal to the population parameter and it is called the *expected value*. Again, this estimate will have some dispersion around the expected value and the standard deviation of the estimate is called the *standard error*. A knowledge of the sampling distribution is essential for assessing the error in the estimate.

Trimmed mean. Sometimes the data on a variable contains some extreme values, which can influence the estimates. For instance, let the average of 50 observations be 6.5. Suppose we add two extremely large or small values to the data and calculate the mean again. The resulting mean could be drastically different from the first mean. Thus, when the data is suspected to have extreme values, one way of finding the mean is to identify such values and exclude them while computing the mean. This is called *trimming* and the resulting mean is called *trimmed mean*. A 10% trimmed mean would be based on data in which the smallest 10% and the largest 10% of the data values are excluded. This method of finding means is necessary in many research problems where certain segments of data appear to be extreme or influential.

1.10 ROLE OF COMPUTER IN STATISTICAL ANALYSIS

A natural question for the researcher is to know how a computer can help in statistical analysis. One has to look into the actual usage of various statistical tools in different areas of research. A large number of advanced statistical tools are available to arrive at valid conclusions from the data collected by the researcher. However, only some primary tools like percentages, means, correlation coefficients and few tests of hypotheses are usually reported in journals. One apparent reason for this is that the calculations require some basic knowledge in arithmetic and lot of patience to carry out them. In social sciences, one may have to handle a large number of related variables, which requires good amount of time and effort to go ahead with statistics. Due to lack of computing facilities (in the past) and insufficient training on the available machines, a compromise has emerged according to which only the basic statistics like means and percentages are reported.

Today, we have a different scenario. Many advanced statistical methods can be applied to various types of data with the help of ready-made computer packages. The researcher has only to load them in the computer and selectively apply the suitable tool for a given situation. Some of the popular statistical software packages are SPSS, SAS, SX, MYSTAT, etc. All these are available in the market. Since India is one of the leaders in software exports, a few Indian companies are marketing statistical software and one package is from Indostat Services, Hyderabad. It contains several statistical routines with special applications to biology, agriculture, econometrics and so on.

REFERENCES

1. Daniel, W. (1983): *Biostatistics—A Foundation for Analysis in Health Sciences*, John Wiley, New York.
2. Rao, C.R. (1993): 'Statistics must have a purpose—the Mahalanobis dictum', *Sankhya*-A, Vol. 55.
3. Snedecor, G.W. and W.G. Cochran (1994): *Statistical Methods*, 8th ed., Affiliated East West Press, New Delhi.
4. Sturges, H.A. (1926): 'The choice of class interval', *Journal of American Statistical Association*, Vol. 21, pp. 65–66.

SUGGESTED READINGS

1. Beri, G.C. (1989): *Marketing Research*, Tata McGraw-Hill, New Delhi.
2. Chandan, S. Jit (1998): *Statistics for Business and Economics*, Vikas Publishing, New Delhi.
3. Indostat Services, Hyderabad, Statistical Software. Contact at indostat@hd1.vsnl.net.in.
4. Reference manuals of statistical package for social sciences (SPSS), SAS, MINITAB and S-PLUS.

DO IT YOURSELF

1.1 Collect data on monthly household electricity consumption in a locality accessible to you. Do you think that it is related to their income levels? Prepare a research plan to study this aspect.

1.2 Collect a list of Indian statisticians and their major research contributions. (*Hint:* see *Glimpses of India's Statistical Heritage*, Wiley Eastern Limited.)

1.3 If you seriously watch cricket on TV, check whether statistics has anything to convey to the viewers?

1.4 Keep track of some daily newspaper and try to make up a database on items like market shares, Government project expenses, weather conditions, etc., for analysis on your PC.

1.5 Prepare a layout for conducting a survey in your institution about the usage of statistical methods in their studies.

1.6 There is a belief among folklore (all over the world) that mentally sick people behave in a peculiar way on full moon days. Prepare a research plan to verify the truth of this belief in the light of a sample data.

Basics of a Computer 2

Not too many years ago, one of the main reasons for grouping data was to simplify the calculation of descriptions such as the mean and standard deviation. With easy access to statistical calculators and computers, this is no longer the case.
— *Richard A. Johnson (1995)*

Statistical analysis starts with the construction of frequency tables and cross-tabulations in every research problem. Depending on the nature of the problem, simple statistics like means and standard deviations may have to be calculated for a large number of variables. Then computer is the only way to do this job effectively. In this chapter, we will understand the basics of a computer and get familiar to work on it.

2.1 THE EVOLUTION OF COMPUTER

The twentieth century has left several remarkable events in human life. Along with great calamities, the Second World War had also witnessed excellent scientific research. Scientists from various fields—statisticians, mathematicians, economists, financial experts and engineers—had worked out on single platform to invent new theories, methods and technologies. Operations research (OR), a subject that is treated as a very powerful interdisciplinary approach to real-world problems, has taken shape during the days of the world war. Statistical methods have come to use in the industrial sector and found applicable in a special area called statistical quality control (SQC). Managers of organizations started appreciating statistical decision-making. All this is for the betterment of human life. The nature of the problem and the technique to be used were so complicated that a machine to perform calculations has become inevitable.

During 1946–52, John von Neumann and his team developed the high-speed digital computer, even though the roots for a calculating machine were laid in the 19th century and even before. Research institutions and universities were the major contributors for the development of computers until 1950, when the manufacturers had started their own research and development in making computers with desired quality at reasonable cost. After several generations in the history of computers, the present *desktop computer* known as *personal computer* (PC) has come into existence. The number of users of computers during the last 20 years has increased drastically, due to the technical advancement in speed, reliability and size reduction of personal computers.

2.2 THE PERSONAL COMPUTER (PC) AND ITS COMPONENTS

The personal computer as we see today is quite different from what it was some 50 years ago. It was big in size, slow and also noisy in its early days. The desktop PC usually appears as shown in Figure 2.1.

Fig. 2.1 Personal computer with its components.

There are four major components in the PC:
  (i) *System:* Here the actual processing of data takes place. It contains several sub-components out of which the central processing unit (CPU) is the most important one.
  (ii) *Keyboard:* Instructions and data to the PC are given by the user through it.
  (iii) *Monitor:* It displays all messages and results.
  (iv) *Printer:* That prints the output on the paper as a hard copy.

We will look into the details of these components in latter sections.
    There are some more fundamental aspects regarding the computer as follows:

Hardware

All the physical devices like monitor, keyboard, mouse, their interconnections, etc., put together is known as *hardware*. The electronic components, the mechanical devices and their functions also belong to this category and the hardware engineer would attend to this job. The user of the PC need not be very much concerned with the details of the hardware, although a minimum knowledge about this aspect would help in proper handling of the machine.

Software

The second important aspect of the PC is the *software*. In simple words, it is a set of instructions to the computer written in a particular code. It is also known as a *program*. A bunch of several related programs is known as a 'software' or a 'package'. Certain software is used for running the PC and it is called as the *system software*. The other software, which is used for specific application like accounting, graphics, inventory, etc., is called the *application software*.

File

Anything we wish to store inside the PC is called a *file*. It could be a 'program' or a text or the data records of several students, etc. Each file is given a name and there will be a large number of files to store. So, we need a procedure to organize these files in a systematic way and access them whenever needed.

Operating system

Every computer works in an environment known by names like disk operating system (DOS), WINDOWS 95/98, UNIX, LINUX and so on. Each one of these is technically called an *operating system* (OS). Leaving the other details, the OS should be understood as a software that tells the user how to organize various files in the PC and also how to manage the information stored in it. It is like an office manager. Without an operating system the PC would be like a person without brain!

 The PC found today in any office usually contains Windows 95 or Windows 98 as the operations system. It would essentially contain DOS as a part of it. Certain software will work only in Windows, while others work in DOS environment only. So, when the PC works in Windows environment, one can work in DOS also. Now, Windows-Me (millennium edition) is also available.

Common application software

The next important aspect is the user-oriented software or application software. Writing a computer program for every problem is time consuming and also not in the reach of all users. It is a specialized job and needs skill in understanding the problem and writing long procedures. Software companies prepare software packages for doing several routine things, without the need to write programs for them. The researcher usually needs a software, for jobs like

  (i) Word processing
  (ii) Data entry and manipulation
  (iii) Performing calculations
  (iv) Preparing graphs and charts
  (iv) Creating slides, animations, etc.

These softwares are used for presentation of data. Now let us look at word processing. There are a number of word processing software available in the market for preparing text like letters, messages, tables, etc. The researcher can use the most commonly available word processor known as WordStar for preparing any text having no mathematical symbols and equations. Many users treat WordStar as a simple and convenient software to type letters, text and numeric data. We will however see latter that another word processor known as MS-Word has more facilities than the WordStar.

The next aspect is data handling, which is a basic requirement in statistical analysis. There is a special software known as dBASE for data entry. There are different versions of this software each having some minor changes and extra facilities. At present, the dBASE software is available in the name of FoxPro or Visual FoxPro. Before this, there was another software known as FoxBASE. Any of these can be used for data entry.

Now the researcher can use (i) FoxPro for all data entry jobs and (ii) WordStar or MS-Word for preparing reports. Several calculations like total, average, variance, count (frequency), percentage, etc., can be done directly with FoxPro. It is also possible to obtain printed output of basic data in any desired form. One can locate for a particular type of data (like a name) among thousands of data records in a few seconds. With simple commands it is also possible to code the data.

The next job for the researcher is preparing graphs, charts, demonstration slides, etc. There is an exclusive package known as MS-Office, which contains Word, Excel, PowerPoint and Access.

  (i) Word is a word processor. This can be used for preparing text with excellent facilities for setting margins and fonts, creation of tables, making text art, typing mathematical equations and so on.
  (ii) Excel is a worksheet or spreadsheet that helps in data entry, manipulation and queries, preparation of tables, creation of graphs and so on. A number of statistical functions can also be performed with Excel.

(iii) PowerPoint is used for the preparation of slides and animation.

(iv) Access is used for handling databases and creation of address labels, mail-merging, etc.

A formal training on each of these software packages would not take much time, but one needs a continuous exposure to the machine and enough work to do on the system.

2.3 PERIPHERAL DEVICES FOR THE COMPUTER

There are certain devices, which are essential for effective functioning of the computer. These are often known as *peripheral devices* and they are briefly discussed below:

Keyboard

The *keyboard* is very similar to that of typewriter but has some additional keys to perform special jobs. These are:

(i) Arrow keys (Home, Page Up, Page Down)

(ii) Function keys (F1 to F12)

(iii) Backspace, Del, Insert keys for text edit

(iv) Shift, Ctrl, Alt to modify the normal key function

(v) Enter key for driving a command or data item into the PC

(vi) A set of numeric keys called numeric pad for quick data entry.

The commonly available keyboard contains 108 to 110 keys.

Mouse

The *mouse* is a special device connected to the PC by a cable. It is a necessary device for working in the Windows environment. The position on the screen can be located easily with the mouse so that any icon on the screen can be activated without using a sequence of keystrokes.

The system

It is the main unit of the PC and usually appears as a vertical box with several gadgets inside. It contains the main processor fitted on a board called the Mother Board. The other important components are the, switch mode power supply (SMPS), the hard disk drive (HDD), the floppy disk drive(s), a compact disk (CD) drive and so on. The other peripheral devices are connected to the system by cables.

Memory

What makes a personal computer different from a calculator? It is the memory of the personal computer, which stores data and instructions in a systematic way and delivers them when needed. It is called the *primary memory*. All that goes into the PC will be stored in this area. It is volatile and gets erased if the power goes off or when the system is switched off. So, a permanent storage method is necessity for storing data. The devices used are external and called *secondary storage devices*. The hard disk and floppy disk are two such devices.

The memory space is measured in terms of units called *bytes*. One kilobyte or simply KB is equal to 1024 bytes and is approximately equivalent to 'half-a-page' of a running text in normal typing mode. One megabyte or 1 MB equals 1024 KB, which is equal to approximately 1,000,000 or one million bytes. This is again equal to approximately 512 pages of a text. Finally, one gigabyte or 1 GB is equal to 1024 MB or approximately 1,000,000,000 bytes or nearly 512,000 pages of a normal text.

The hard disk drive

The *hard disk drive* (HDD) is the secondary storage unit fitted inside the system. Under normal conditions the user will have no access to remove it from the system. The capacity of a hard disk is in the range of 10 GB to 20 GB and even more. The HDD has a built-in facility for read–write operations so that the user can save the data and other files in the HDD while working on the system.

The floppy disk drive

A *floppy disk* is a small portable medium of data storage. The floppy disk available today is called '$3^1/_2$ inches disk' and its storage capacity is 1.44 MB. We can use the floppy to make copies of the files currently maintained in the hard disk. In case of failure of the hard disk it is possible to use the floppy disk and copy the contents into the hard disk. We need a device called floppy disk drive (FDD) to carry out read–write operations with a floppy disk. A few years ago there were big floppy disks of $5^1/_4$ inches with a corresponding driver to handle them. They are no more in use except where there is a drive of that type.

Monitor

The output from the computer can be viewed on the monitor. The text we type, the results of a calculation and the error messages, if any, are seen on the monitor. We can use a monochrome (black and white) monitor or a colour monitor. For Windows usage, a colour monitor is suggested. The quality of the output depends on the technical aspects of the monitor. The monitor plays a dominant role in viewing the output before making a presentation on slides and transparencies.

Printer

In order to have a hard copy of the output on a paper we need a *printer*. There are different types of printers: dot matrix, ink jet and laser printers. The dot matrix printer is commonly found in offices, shops, railway booking counters and even in banks. It is used for such purposes where high quality output is not the criterion but a large volume of prototype work is to be done. Printing of a railway ticket, printing a monthly salary statement are examples of this type of work. Dot matrix printers are available in two modes namely (i) 9-pin mode and (ii) 24-pin mode. The 24-pin printer gives a higher quality output than the 9-pin printer.

The width of the printer determines the type of output that can be taken on it. There are two types again. The first one is the 80-column printer, which is used for typing text, and data records in which the number of characters (including blank spaces) does not exceed 80. For some business applications and also in accounting problems, we may have to use a paper of width longer than 80 columns and for such applications the 132-column printer is used. This printer can be used to load the 80-column paper also but the other way is not possible.

The paper we use can be a single sheet or a continuous paper, which will be automatically fed to the printer with the help of a pull mechanism. We can use paper that supports multiple copies usually 2 or 3 known as 2-part or 3-part paper, which comes along with a carbon paper, sandwiched between sheets.

The dot matrix paper has a low maintenance cost because the output is obtained through a ribbon *cartridge*, which can be replaced at a low cost. The ink jet and laser printers are known for their high quality output. They are essential when high quality output and graphics are to be printed on paper. They are available in standard sizes with a provision for colour printing also. The printing is based on a special 'refill cartridge'. The maintenance is however costlier than that of dot matrix printer. For printing images like photos, a laser printer is the best one to use.

CD-ROM drive

A floppy disk is a medium on which we can *write* programs or text through the PC. We can also *read* the contents of a file copied from other sources. Thus, a floppy can be used for both reading and writing.

A different medium of data 'storage and transfer' is a disk on which information is stored by a special device which the user cannot modify once written on it. We can only read the information or copy it on to the hard disk of the PC. This is certainly a better way of communication than the floppy disk. This is called a *read-only memory* (ROM) and the disk is called the *compact disk* (CD). Such disks are today known as CD-ROM disks or simply CDs. However, there are rewritable CD, which can be used for modifying the contents with the help of a device called CD writer.

Things like software packages, lectures, historical information on science and technology, etc., are nowadays available only on CDs. The user cannot change the contents so that the originality is maintained. Most companies release their software on CDs only. Popular computer magazines (and books) supply a CD with some of their issues and a lot of information is contained in it. The CD is a robust medium unlike a floppy. Its storage capacity is many times larger than that of a floppy.

In order to read the contents of a CD we need a special drive similar to a floppy drive. This is called the CD-ROM drive. The speed at which it reads the contents of a disk is an important parameter and measured as 32×, 40× etc. It is suggestive to have a CD drive fixed to the PC. While operating with Windows, if any files get corrupted, we have to copy them from the original CD and this requires the CD drive. For the researcher, a lot of information can be obtained on CDs. The researcher can record all his findings in the form of data, graphics, diagrams, photos, etc., and can even release a CD for use by others.

Multimedia

A PC is today not only a computing machine but also a high quality information and communication medium. In addition to viewing photos and text on the monitor, we can listen to expert lectures, special music, field demonstrations and so on through a facility called *multimedia*. It is a facility in the computer and is a combination of many softwares that supports audio and video at a time. A set of devices known as multimedia kit is available that contains a CD-ROM drive, speakers and a microphone along with special cables to carry audio and video signals.

As a result, we can listen to voice while viewing a clipping. This facility has become a common requirement to the researcher, for understanding details about scientific investigations. We can also speak before the microphone and our voice goes along with our demonstrations. The same can be copied to a floppy or to a CD.

Modem, telephone and the Internet

With the help of the PC today it is possible to connect ourselves with the users in various parts of the world using a facility called the *Internet*. The Windows 98 operating system provides built-in software for Internet connectivity. The PC has to be connected to the telephone through a gadget called *modem*. By registering as an Internet user, we can connect our PC to the Internet. Information on various aspects will be available through distinct sources called *Web sites*.

Through Internet we can use a very convenient mailing facility, called the *e-mail*. By creating an e-mail address we can communicate with others very quickly. For researchers, it is the best method of communicating research articles, seeking clarifications from outside experts and so on, in short time. Thus, the Internet is a necessity than a comfort for the researcher.

2.4 GETTING STARTED WITH THE PC

When the system is switched on, the operating system inside will carry out certain checks and if everything is well, either the message 'Loading MS-DOS' or the message 'Starting Windows' will appear on the screen of the monitor. This stage is called *booting*.

While working with the PC the monitor will display the source of data from which the current work is going on. In DOS we find the symbol C:\> to indicate that we are working from the hard disk. This letter 'C' is conventionally reserved to designate the HDD. By the same logic, the letter 'A' is used to designate the FDD. The Windows system has also the same way of displaying the drives.

When the system is booted from DOS, we get C:\> symbol directly. If it is booted from Windows we get a menu with several icons for different jobs and we click on the MS-DOS icon from the toolbars by using the mouse. We then get C:\WINDOWS> symbol. It means we are in the Windows folder and we can change to any other folder from there.

With MS-DOS we find the command line as C:\> with a blinking character called the DOS prompt. It means that we are in the main directory of the hard disk. A *directory* is collection of files. It is a good practice to create a directory for each user and for each purpose so that all the files common to a particular job or to a particular user will be in a specific directory. It helps in locating the files easily.

2.5 CREATING A DIRECTORY (FOLDER)

A *directory* is a collection of files stored at a particular location on the hard disk or on the floppy disk. Every file shall have a name and the directory is also given a name. The file name contains two parts. The fist one is the *name* and the other is called the *extension* and they are separated by '.' in between. Examples of file names are MARKS.DBF, SVU.DOC, STAT.CHI, etc. The software usually allots the extension itself. The name should have a maximum of 8 characters and the first character shall be a *letter* and not a numeral. Special symbols are not allowed while defining a file name.

All the files in the hard disk are stored in a master directory called *root directory* or *main directory* denoted by C:\>. If we do not specify the directory to store a file, it is automatically *saved* in the root directory. It is however, not a good practice to save everything in this directory. It is like writing all notes in a single large book and latter suffer to locate a particular item from it.

The second purpose of having a separate directory is not to mix up the data files with other software files. This is because, if a file becomes 'corrupt' due to some *virus* the same file may affect the software and the whole thing becomes useless.

So, we need a separate directory or a sub-directory for each purpose. In Windows 95 or 98, there is a built-in directory called 'My Documents' and it is the *default directory*. In Windows, a directory is shown as a FOLDER. If we do not create a directory to store files, the PC automatically stores it in the folder 'My Documents'. Suppose the researcher wants to create a directory named as STATMAN. All files can then be stored in this directory. Similarly, one can use a directory name like PROJECT, HOSTEL, etc., indicating the purpose of the directory.

To create a directory in MS-DOS we proceed as follows:

- Switch on the PC and wait for a few seconds until the C:\> prompt appears.
- Now type the command in one of the following ways.
- Type MD\STATMAN or md/statman or MD STATMAN and press the Enter key. Then the directory gets created (MD stands for Make Directory) as a sub directory to the main directory.
- Now type CD\STATMAN followed by Enter and the control goes from C:\ to C:\STATMAN. We get the screen as C:\STATMAN> (CD stands for change directory).

(At the end of each command it is necessary to press the Enter key without which the command does not reach the system.)

We can view the contents of a directory by typing DIR (means directory) followed by Enter. Optionally, we can use DIR/P for a page-wise list of files or DIR/O/P for an alphabetically ordered page-wise list or DIR/W for a horizontal look.

We find no files except two invisible files in the directory of STATMAN. Now all is well with our system in DOS and we are working in the current directory named STATMAN.

If we have booted from Windows we have to choose the Explorer icon or the My Computer icon and click on it with mouse. On the File menu we find an item called NEW. By clicking on it we can create a New Folder and we simply type STATMAN on the New Folder at the blinking point. This creates the directory. These details are also discussed in Chapter 4.

2.6 GETTING ON TO FOXPRO

Most of the statistical data should be entered using a numerical code rather than an alphabetic code. FoxPro is a convenient package for data entry and manipulation. FoxPro is available in both DOS and Windows mode. The DOS version is known as FoxPro 2.5 and the discussion in this chapter is based on this version. We can also work in FoxPro 2.6 or Visual FoxPro, which works in Windows environment. It should be noted that earlier versions like dBASE III or FoxBASE can also be used in DOS version. Assume that FoxPro 2.5 is installed in the PC. With proper installation and changing the *path* in the AUTOEXEC.BAT file we can access FoxPro from any directory. For instance if we use CD\STATMAN we get on the screen C:\STATMAN>. Now if we type 'fox' followed by Enter (capital or small letters make no difference), we get an opening message about FoxPro and a full screen appears with the menu on the top as shown in Figure 2.2.

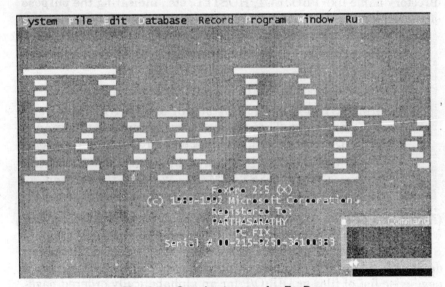

Fig. 2.2 Opening screen for FoxPro.

A small window called the *command window* appears at the bottom-right corner of the screen (we can change this position if we like). All FoxPro commands are to be typed in this window each followed by Enter.

An interesting feature of FoxPro commands is that the *first four letters* of the command are enough to work. We need not type the full command. For instance a command like BROWSE can be typed as BROW and it works. Similarly the command REPLACE can be used as REPL.

Now type the command SET STATUS ON or SET STAT ON. We get the status line at the bottom of the screen, which indicates the current file name and the number of records of the file. We can also type SET CLOCK ON and the clock is displayed on the top right corner of the screen.

A great help about the usage of various commands can be found by pressing the function key F1 or by typing HELP in the command box.

To get out of FoxPro we have to type QUIT and press Enter key. The system goes to C:\> and we can shut down the PC.

As an alternative to typing the commands, we can select them from the main menu by using Alt key. For instance, Alt + F gives a small menu called system commands as shown in Figure 2.3.

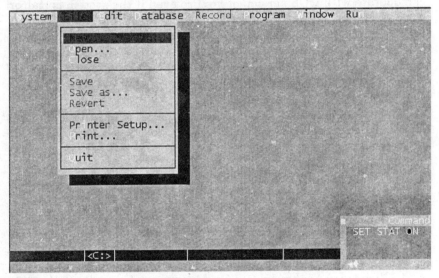

Fig. 2.3 File menu in FoxPro.

We can select one of the items like Open, Save, etc., with the help of arrow keys and press Enter key. The *left-click* of the mouse is used for actions and the *right-click* is for displaying the properties. If we have booted from Windows, we can use the mouse to click on various items. To shut down the system while in DOS, type CD\ and press Enter. Then switch off the system.

2.7 DOS OR WINDOWS?

Windows operating system is undoubtedly more convenient than DOS for the user. However, a regular user of DOS finds a difference in booting time, method of creating directories and looking at directories, etc., when compared with Windows. For one who can type commands at a good speed, the DOS commands are found easier to use than selecting the commands with mouse in the Windows system.

Suppose a few lines of text like a note is to be typed on a paper. For those who work in DOS environment, WordStar is convenient because its access time is significantly lower than the time needed to open MS-Office and Word from there. A small amount of random access memory (RAM) like 4 MB or even less would be enough to operate a DOS-based package like WordStar. The Windows-based software however requires a higher RAM and a high-speed processor to display the graphics. In the DOS system, data entry can be done with FoxBASE, which is much similar to FoxPro except for graphic utilities. Thus, DOS has its own advantages for the user. The user can operate with DOS version of FoxPro even if the system is opened with Windows.

We can bypass the Windows booting by pressing the function key F8 during booting when the system details are being displayed. A menu appears from which we can select the option *Command Prompt Only* and press Enter key. This will bypass the Windows files and the system works in DOS only.

Windows is however the only way to deal with advanced methods of data and text processing, graphics, Internet, etc., which are the basic needs of today's researcher. We will focus on the details of Windows in Chapter 4. In the following chapter we look into some details regarding the creation of files in FoxPro and doing some statistics using it.

REFERENCES

1. Richard A. Johnson (1995): *Miller and Freund's Probability and Statistics for Engineers*, 5th ed., Prentice-Hall of India, New Delhi.

2. Taxali, R.K. (1996): *FoxPro 2.5 Made Simple for DOS and Windows*, BPB Publications, New Delhi.

DO IT YOURSELF

2.1 Visit a computer lab in your institution and examine which peripheral devices they have.

2.2 What is write protecting of a floppy disk? Try to open and close it.

2.3 Create a directory in your name and try to change over it from another directory.

2.4 Use DIR and press Enter key. Try to read out what file are available on your PC.

2.5 Use each of the following commands one at a time. Press Enter key after each command. See what happens.

DIR/W   DIR/O   DIR/O/P   DIR > PRN
CLS   DATE   TIME

2.6 Check whether your PC has FoxPro, FoxBASE or dBASE III. If not, contact your software consultant and get it loaded.

2.7 Open FoxPro and examine different items shown on the top of the screen. Use Alt key for this purpose.

Data Handling and Statistics through FoxPro   3

With simple commands, FoxPro can perform some basic statistical calculations. In this chapter, we understand the job of creating and editing a data file for a given situation and then examining the statistical aspects of the data. All the discussion will be held in DOS environment only. If the system is already working in Windows (as done in many computers) we simply click on the short cut toolbar for MS-DOS.

3.1 HOW TO CREATE A DATA FILE?

In order to create a data file, we shall first choose a name to that file. Assume that our file is to be named as TEMP (to mean temporary) and that we wish to keep this file in the directory STATMAN. As mentioned in the previous chapter, we can open FoxPro from any directory. So first go to C:\STATMAN and then type FOX.

Every database file is a collection of *records* and every record contains *fields*. Every field is given a name like SNO, AGE, SEX, etc. No two fields shall have the same name. Even if we use the same name FoxPro would warn about it and we change the name. Data on all the fields pertaining to one case or one questionnaire or one subject of study makes up a *record*.

The way in which fields of a file are defined is called the *structure*. FoxPro supports following types of fields:

(a) Numeric fields to store numbers

(b) Character fields to store names and labels

(c) Date fields to store dates and

(d) Memo fields to store long titles or a few lines of a text.

There are some more types of fields meant for other applications. The following is the method for creating a data file:

- Type CREA TEMP in the command window and press the Enter key.
- We find a screen as in Figure 3.1.

Type the name of the field and press the Enter key. The cursor moves to the next heading. Press 'N' for numeric data and 'C' for character data and define the width of the field in terms of the number of spaces required to store that data. For numeric fields, the decimal places can also be specified. While defining the width for numeric field, we have to count the total length including the decimal places and type it as width. Thus, if we need 2 decimal positions type '2' under DEC.

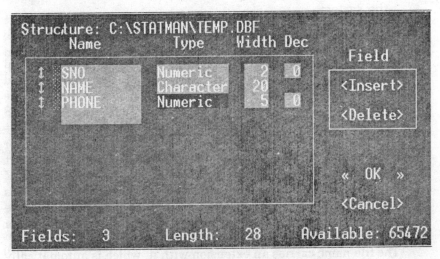

Fig. 3.1 FoxPro screen for creating a new file.

- In order to enter the data like 354.63, the width must be 6 including the decimal point and the decimal places shall be 2. The advantage of defining the field width is that we need not type the decimal point while entering the data. Just by typing 35463, it goes as 354.63. This would save a lot of time while handling large amounts of data.
- If all the fields are defined and if we are satisfied with the structure we have to save the structure by pressing Ctrl + W or by choosing the OK button on the screen and clicking on it with mouse. We can also use Tab and press Enter key at the OK button.
- Now the following message appears on the screen.

INPUT DATA RECORDS NOW? (YES/NO)

- Select Yes button to input data immediately or No button to do it latter. If Yes, we get one blank record of all fields in the given order and we can go on entering the data.
- The length of the record is (2 + 20 + 5) + 1 = 28. One extra space is provided for each record to put '*' mark, in case the record is deleted (see Section 3.2.2).

- After entering a few or all records, press Ctrl + W to save the data. Now we can close the file and continue working on it latter.
- To close the file type CLOSE ALL
- To leave FoxPro type QUIT
- This will bring us out of FoxPro. We can shut down the PC, if we wish, by typing CD\ followed by Enter and then switch off the system.

We can delete the file TEMP in two ways. In the command window, type the command DELE FILE TEMP.DBF. Alternatively, we can type QUIT and leave FoxPro. This takes us to the directory STATMAN. Then we can type DEL TEMP.DBF and get the file deleted.

3.2 DATA FILE FOR THE TRIBAL FOOD PROBLEM

The data on the tribal food problem mentioned in Chaper 1 is a part of the study carried out by Lavanya (1999). It contains records of 150 persons with 50 persons under each of the three tribal categories namely Sugali, Yanadi and Yerukala. In each category there are again males and females. Let us create a file in FoxPro for this data and name it as TRIBAL.

The file name carries an extension with it, which is automatically given by the software. The extension for the data file is '.DBF'. For instance the file TRIBAL appears in the directory as TRIBAL.DBF. All FoxPro data files appear with this extension. Technically speaking FoxPro, dBASE III, dBASE IV and FoxBASE files are all treated as dbase files.

FoxPro creates an automatic back-up file to each DBF file with the extension '.BAK'. In case of any damage to the DBF file we can use the BAK file and carry out the work. If we type DIR in the command window, we see the list of all dBASE files. The command DIR *.BAK will display all the back-up files.

In order to take the data to the computer the following data structure has been created in FoxPro.

1. SNO   (Serial Number)
2. SEX   (Male 1, Female 2)
3. AGE   (Years)
4. HT   (Height in cm)
5. WT   (Weight in kg)
6. BMI   (Body mass index in kg/m^2)
7. BMR   (Basal metabolic rate in kcal/day)
8. PRO   (Protein in gm)

9. FAT   (Fat in gm)
10. CHO   (Carbohydrate in gm)
11. CAL   (Calcium in mg)
12. IRON   (Iron in mg)
13. VITA   (Vitamin-A in µg)
14. VITC   (Vitamin-C in mg)
15. EIN   (Total energy intake in kcal/day)
16. EEX   (Total energy expenditure in kcal/day)
17. EBAL   (Energy balance in kcal/day)
18. CODE   (Sugali 1, Yanadi 2, Yerukala 3)

The record corresponding to each person now contains these 18 fields. (The complete file of 150 records with these fields is given in Appendix B. The reader is expected to create this file and proceed.) A sample data of one person to be stored as one record is as follows:

SNO = 1	BMR = 1459.3	VITA = 2321
SEX = MALE	PRO = 54	VITC = 31
AGE = 21	FAT = 37	EIN = 2654
HT = 158	CHO = 518	EEX = 3674.1
WT = 51	CAL = 360	EBAL = −1020.1
BMI = 20.43	IRON = 21	CODE = SUGALI

In order to create a file we proceed as follows:
- Switch on the PC and go to the directory STATMAN
- Type FOX
- Type SET STAT ON
- Type CREATE TRIBAL and press Enter
- Type the STRUCTURE as per the fields given above
- Enter one or two records and save the data by using Ctrl + W.

We can as well continue typing until all the records are over. Suppose we have typed a few records and wish to close the file for some time. Then, type CLOSE ALL to close the file and then type QUIT.

- To continue typing the data again, type FOX
- Type USE TRIBAL
- Type BROWSE (or simply BROW) to view the records of the file on the screen
- The first 14 records of the file appear as shown in Figure 3.2
- All the 18 fields may not be visible on the screen. Use Enter key to move to the remaining fields
- Press the Esc key given on the top left of the keyboard. This will take us to the command window

- Now type APPEND or APPE to add new records
- The screen appears with all the fields of each record along with a blank provision against each field, for typing data.

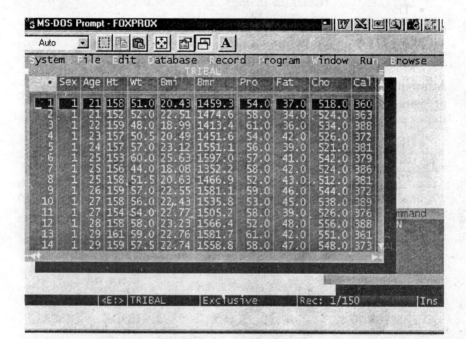

Fig. 3.2 FoxPro screen after the BROWSE command.

After typing all records (or a few records, if desired) save the file by pressing Ctrl + W. By using the command BROW the data appears as a table. The number of records in the file and the number of the current record will be displayed in the *status line* at the bottom of the screen. Suppose we are at the beginning of the file and there are 150 records in the file TRIBAL, then the status line would display the file name as TRIBAL and we find the number 1/150 displayed on the status line. It shows that we are in the first record of the file, out of 150 records. This is the advantage of the command SET STAT ON.

By using the arrow keys, we can move to any record and make changes, if necessary.

3.2.1 Changing the Structure

If we wish to create a new field or delete an existing field from the structure, it is done by the command MODI STRUCTURE.

- A field can be inserted before another field by pressing Ctrl + I. We get a blank field with name NEWFIELD

DATA HANDLING AND STATISTICS THROUGH FoxPro

- In that field, the name and the type can be entered
- An existing field can be deleted by pressing Ctrl + D
- The type of field can be changed from character to numeric by typing 'N' or 'C' as required
- The field width can be changed by re-typing on the existing field width.

After making any change, it is necessary to save the structure by pressing Ctrl + W. We can also carry out editing by selecting the items in the top menu by using mouse or Alt key.

- By typing the command DISPLAY STRUCTURE we can see the structure of this file as shown below. The total width is 73 + 1 including the position of '*' mark.

```
Structure for database : C:\STATMAN\TRIBAL.DBF
Number of data records : 150
Date of last update    : 30/9/99
Field   Field Name   Type        Width   Dec
  1       SNO        Numeric       3
  2       SEX        Numeric       1
  3       AGE        Numeric       2
  4       HT         Numeric       3
  5       WT         Numeric       4       1
  6       BMI        Numeric       5       2
  7       BMR        Numeric       6       1
  8       PRO        Numeric       5       1
  9       FAT        Numeric       5       1
 10       CHO        Numeric       6       1
 11       CAL        Numeric       3
 12       IRON       Numeric       2
 13       VITA       Numeric       4
 14       VITC       Numeric       2
 15       EIN        Numeric       6       1
 16       EEX        Numeric       6       1
 17       EBAL       Numeric       8       1
 18       CODE       Numeric       2
**        Total **                74
```

- We can also get a print out of this structure by typing the command DISPLAY STRUCTURE TO PRINT
- As a short-cut we can use DISP STRU in place of DISPLAY STRUCTURE.

When there are many fields in the file, it is a good practice to keep a print out of the structure and make changes, if any, in the fields and their types.

3.2.2 Editing the Data File

We can edit the data whenever required. By any reason if we have typed incorrect data or missed some records we can rectify such errors by editing the file. We can also carry out structural changes like creating new fields or deleting fields in a file.

The command to edit data is EDIT. This will show the full record and we can edit any field in it. To move from one field to another we can use the arrow keys of the keyboard.

As an option, we can edit selected fields only. For instance, the command EDIT FIELD SEX, AGE shows only the two fields namely SEX and AGE for editing. In this command, the field names should be separated by comma only.

There is one chief advantage by using FoxPro for data entry. Suppose we have a field whose width is 1. An example is SEX in the TRIBAL case. While editing a record, the moment we enter 1 (or 2), the control automatically shifts to the next record and we need not press the Enter key.

At some stage, a data value may require a wider space like 3 characters for which we might have earlier given only 2 characters as width while creating the file. Then we have to type the command MODI STRUCTURE. This shows the structure on the screen from which we select the field and change the field width. We then save the modified structure by using Ctrl + W.

The menu on the top of the screen will give all help for editing the data and we can select the items by using either Mouse or the Tab key.

To delete one record from the file, it is a simple job as under:

- Go to the desired record say 10th record by typing 10 in the command window
- Type BROW
- The 10th record appears on the screen
- Now type Ctrl + T
- Then an asterisk (*) appears against the record on the left side
- It means the 10th record is deleted.

We can also get this effect by selecting the buttons RECORD ▸ DELETE from the main menu (▸ denotes followed by). We can also delete a group of records like all records with numbers 51 to 65 as follows.

- Go to 51 by typing 51 in the command window
- Type DELE NEXT 15 (DELE is a shortcut for DELETE).

Suppose we wish to delete all the remaining records after 50. To do this we first go to record 51. Then type the command DELE REST. All the records are deleted and contain an asterisk (*) mark against them.

Suppose a file contains 1000 records and we wish to divide the work to two or more persons for data entry. One way of doing this is to create the structure in one PC and copy it on to the floppy disk. This file can then be loaded into other computers. Suppose the work is assigned to 2 separate PCs with file names SVU1 and SVU2 respectively with 500 records in each file. After entering the first 500 records in SVU1 we can *append* the records of SVU2 to SVU1 with the following sequence of commands.

USE SVU1
APPEND FROM SVU2
BROW

This will create 500 + 500 = 1000 records in SVU1. Note that, SVU2 still contains 500 records. If not required, it may be deleted with the command DELE FILE SVU2.DBF. A message appears that the file has been deleted. Some times we may not get such message. Then type the command SET TALK ON and then type the previous command. Then the message will be displayed.

When the records are deleted, they are not permanently removed from the file but labeled with a '*' mark. For all calculations these deleted records are ignored even though they appear when the file is browsed. To delete them permanently the command is PACK.

A deleted record can be recalled from the file before the file is packed. The command is RECALL or RECALL 10 to recall the 10th record.

Sometimes a particular record may have to be inserted in a file before another record. For instance, we may wish to insert a new record before the 4th record. To do this, type 4 in the command window and press Enter. Then type the command INSERT BEFORE and press Enter. A new blank record appears before the 4th record. We can now type data in it.

Suppose we wish to add a new variable named as EBAL to mean energy-balance. To do this, type MODI STRU and add a field EBAL at the end. Then press Ctrl + W. Now type the command

REPLACE ALL EBAL WITH (EIN-EEX)

and press Enter key. All EBAL will be replaced with this difference. All these changes can be seen on the screen with the BROW command.

We can sort the file in alphabetic order or numerically increasing order of a field by using the command SORT. The usage is SORT ON {field name} TO {new file name}. As an example, we can apply the following steps:

USE TRIBAL
SORT ON SEX TO SSS
USE SSS
BROW

This would display the file in such a way that all records for MALES appear first followed by all the records for FEMALES. The disadvantage is that whenever a file is sorted on a variable, a new file is created which takes a significant amount of disk space, for large files. So, we have to delete such a file soon after the purpose is over.

As an alternative, we can use the command INDEX to create the effect of SORT without creating a sorted file. Indexed files are very useful while handling multiple database files. (More details can be had from any book on FoxPro.)

3.2.3 Printing the Data File

If we use the command LIST (without TO PRINT), the results appear only on the screen and not on the printer. The data stored in the file can be printed as a hard copy by using the clause TO PRINT, provided a printer is available and connected to the PC. The following are some useful commands for controlling the view as well as printing of records on a printer:

  LIST TO PRINT

This would print all the records of the file on the printer. We have to ensure that the printer is ON and the paper is loaded. We can either use a continuous paper or single sheet, if desired. The print appears along with a special field, called Record Number, which is usually not required. We can avoid printing the record number by using the command

  LIST OFF TO PRINT

If there are many long fields or if the field names are long, the length of the record may exceed 80 characters, and each record will be printed in more than one line. It may then be difficult to read the contents. So, there is a provision to print only the desired fields of the file by using the following statement:

  LIST FIEL SNO, SEX, CODE, BMI TO PRINT

This would print only these four fields for ALL the records. (FIEL is the shortcut for FIELD.) We can also print only a few records say the first 20 records by using the command

  LIST NEXT 20 FIEL SNO, SEX, CODE, BMI TO PRINT

Before we apply this command, we shall go to the first record so that next 20 records can be listed. To go to the first record we have to type '1' and press Enter or type GOTOP and press Enter.

The width of the paper is usually 80 columns or 132 columns. Depending on the printer available, we select the paper width. Normal printing work can be done with 80-column paper. We can load an 80-column paper on a 132-column printer also but not the other way. If the record length exceeds 80 characters (including blank spaces), we can choose a 132-column paper to print the data. We can sometimes use the 80-column paper and select the output fields in such a way that they appear all in one line. It however depends on the number of fields and their width. For instance, consider the command

LIST OFF FIEL SNO, AGE, SEX, BMI, BMR TO PRINT

This would print all the records for these selected fields only. In the next phase, we can select the remaining fields and print all the records. Suppose we use the command

LIST OFF NEXT 50 FIEL SNO, AGE, SEX, BMI, BMR TO PRINT

This will print the first 50 records with the selected fields. After this printing is over, type 51 in the command window and reuse the above command. Then the records 51 to 100 will be printed. This is one way of controlling the output while printing the data.

Blank entry in a field

One important issue in the data entry is that of handling a *blank* in a data field. Suppose the BMI value is *not available* for a few records. If we leave a blank for that field in that record, the result is taken as '0'. This is because all blanks in a numeric field are automatically replaced with '0'. Then this is not a correct entry. So the only way is to create a *character field* instead of *numeric field* and mark NA to indicate 'not available'. We can also specify a blank space with a blank or a '-' mark in a character field.

Value command

The contents of a character field can be addressed by FoxPro in a different way by giving the value in quotes. For instance, if BMI is defined as a character field, the command LIST FOR BMI > 25.0 will show an error message while LIST FOR BMI > '25.0' is accepted. Thus numeric comparisons and calculations with character fields are handled in a different way. The command is VAL( ). We can see help in FoxPro and get the method of using this command.

3.2.4 Statistical Calculations with FoxPro

It is now time to look at some commands that help in getting the basic statistics like count, mean, variance, standard deviation, etc., with the help of FoxPro. Let us look at this aspect with the help of the file TRIBAL.DBF.

- USE TRIBAL and BROWSE
- Press Esc key to get a blank screen except the command window.

We now perform some statistical calculations using the following commands. The result of each command is also shown.

(i) CALC AVG(BMI): This gives the average of all the values of BMI and the result is 19.93. We can also use the simple command AVERAGE (BMI) and press Enter key.

(ii) CALC CNT( ) FOR SEX = 1: This gives the number of cases in the file corresponding to SEX = 1 (Males). The output is 75.

(iii) CALC CNT ( ) FOR SEX = 2 AND CODE = 2: This gives the COUNT of the number of cases corresponding to female Yanadi respondents of the study. The output is 25.

(iv) CALC CNT( ) FOR EBAL > 0: This gives the count of all cases for which EBAL is positive. The output is 1. It means there is only one case with positive energy balance. We can view this record by using the command LIST FOR EBAL > 0.

(v) CALC MAX(BMI): The maximum value of BMI is given by this command and the output is 26.37.

(vi) CALC MAX(BMI) FOR SEX = 1: The maximum of BMI among males is given by this command and the output is 25.63.

(vii) CALC MIN(BMI) FOR SEX = 1: The smallest value of the variable BMI is given by this command as 17.70.

(viii) CALC STD(BMI): This gives the STANDARD DEVIATION of BMI over all the 150 cases and the output is 2.11.

(ix) CALC STD(BMI) FOR CODE = 1: The standard deviation of BMI among only Sugali subjects is given by this command and the output is 2.27.

(x) CALC VAR(BMI): The variance of BMI is given by this command and the output is 5.15.

(xi) CALC SUM(BMI) FOR CODE = 3 AND SEX = 2: The sum of all BMI for Yerukala females is given by this command and the output is 492.20.

We can also use multiple commands in one statement like

CALC AVG(BMI),STD(EBAL),VAR(BMI),SUM(BMI)

This gives all the results at a time (try it!).

There are many other mathematical functions for which the answer can be obtained directly. For more details look into Help and read the topics from it.

3.2.5 Frequency Tables and Cross-tabulation

Simple FoxPro commands can be used to find out the frequency distribution of the data values. In order to get a full and printable output we however need some knowledge of programming in FoxPro. Let us get the frequency table of BMI given in the file TRIBAL. Following are the steps:

- CALC MIN (BMI). This gives 14.60
- CALC MAXIMUM (BMI). This gives 26.37
- Let us take classes as 14–16, 16–18, 18–20, 20–22, 22–24, 24–26 and 26–28 and count the number of cases falling in each class
- COUNT FOR BMI >= 14 AND BMI < 16. This gives 4 records in this interval. So, the frequency (count) is 4
- COUNT FOR BMI >= 16 AND BMI < 18. This gives 19 records in this interval.

Like this, all the frequencies can be counted and filled in the table as follows:

BMI range	Number of cases
14–16	4
16–18	19
18–20	57
20–22	47
22–24	16
24–26	6
26–28	1

The total frequency is the number of records in the file and this requires no calculation. In our example, it is 150. We can also get this number by typing the command '? RECCOUNT( )' followed by Enter.

Let us divide the values of AGE into two groups according as AGE <= 25 and AGE > 25.

The researcher may wish to make a cross-tabulation of cases falling under these two groups with respect to SEX. Since the variable SEX is at two levels and AGE is at two levels, we get a 2 × 2 table. The following commands will give the four frequencies of the table.

COUNT FOR SEX = 1 AND AGE <= 25   (This gives 13)
COUNT FOR SEX = 1 AND AGE > 25   (This gives 62)
COUNT FOR SEX = 2 AND AGE <= 25   (This gives 34)
COUNT FOR SEX = 2 AND AGE > 25   (This gives 41)

These numbers can be filled in the form of a table as shown below:

	AGE <= 25	AGE > 25
MALE	13	62
FEMALE	34	41

We may expect the computer to do this automatically without using the above 4 commands. It is indeed possible by writing a small program in FoxPro. In Chapter 10, a few programs are given.

3.3 FOXPRO FILE FOR BLOOD-BANK DATA — A CASE OF HOSPITAL MANAGEMENT

Vijaya Kumar et al. (1997) have studied some aspects of the pattern of blood donations received at the blood bank of a hospital. The data for the blood bank is available for different months of the years 1993 to 1996 as shown in Table 3.1. Data was available only from March 1993. Depending on the type of questions to be answered from this data, we have to create the structure for the data file. Suppose we wish to compare the total number of donations (i) in each month of the same year, (ii) in different years for the same month, (iii) according to each blood group and so on. We can answer these questions directly from FoxPro.

Let us create a FoxPro file by name BLOOD for this data and keep it in the directory C:\STATMAN. As done in Section 3.2, let us first go to the directory C:\STATMAN. Then open FoxPro. In the command window type CREA BLOOD and press Enter key.

Table 3.1 Monthly Collections of Different Blood Groups

Month	Blood groups							
	1993							
	O-POS	O-NEG	A-POS	A-NEG	B-POS	B-NEG	AB-POS	AB-NEG
MAR	8	1	4	0	6	1	0	0
APR	14	2	0	0	8	0	0	1
MAY	28	2	6	4	4	0	1	0
JUN	35	0	4	0	7	0	1	0
JUL	30	0	19	0	26	3	5	0
AUG	4	2	1	2	9	0	0	0
SEP	42	0	20	0	14	0	5	1
OCT	58	1	29	0	67	3	5	0
NOV	69	19	41	0	39	5	8	1
DEC	93	3	11	0	47	7	6	1

Table 3.1 (contd.)

				1994				
JAN	60	0	39	4	46	1	13	4
FEB	94	6	20	0	71	6	4	0
MAR	131	18	40	4	79	5	15	3
APR	97	12	49	6	54	5	16	4
MAY	119	3	39	3	75	0	13	2
JUN	161	5	45	6	114	10	11	1
JUL	113	14	43	7	75	16	18	1
AUG	113	12	50	5	80	4	20	3
SEP	109	6	41	3	69	9	13	0
OCT	85	6	34	3	56	2	11	0
NOV	75	4	37	4	73	6	12	3
DEC	78	3	33	5	48	6	16	1
				1995				
JAN	131	8	53	4	93	5	10	1
FEB	94	7	54	1	51	8	19	3
MAR	102	13	45	1	76	6	12	1
APR	82	4	44	1	54	3	13	0
MAY	98	5	33	6	59	6	5	11
JUN	83	2	35	3	40	4	9	2
JUL	73	2	21	3	31	8	20	0
AUG	96	12	40	9	59	9	20	2
SEP	117	11	46	9	68	13	15	0
OCT	95	13	55	6	72	7	12	0
NOV	75	10	37	4	75	4	12	0
DEC	89	7	39	1	54	9	7	1
				1996				
JAN	68	7	42	4	43	7	12	1
FEB	130	14	40	5	78	6	20	3
MAR	100	11	43	2	65	5	18	2
APR	84	8	28	3	76	11	18	2
MAY	133	21	57	5	94	3	9	3
JUN	58	5	22	3	47	2	7	3
JUL	94	10	56	8	62	4	12	2
AUG	58	3	20	9	44	3	10	1
SEP	91	9	35	2	60	7	8	0
OCT	87	7	64	1	71	5	10	0
NOV	80	4	44	5	63	7	10	2
DEC	47	6	26	2	50	3	4	1

Let us define the fields as SNO, YEAR, MONTH, OPOS, ONEG, APOS, ANEG, BPOS, BNEG ABPOS, ABNEG as the basic fields for which we can directly enter the data. OPOS indicates O-POSITIVE group and ONEG indicates O-NEGATIVE group. The other groups are similarly coded.

All the fields are numeric except MONTH. The field width is given in such a way to accommodate the maximum number of digits.

If we are working with FoxPro for Windows, we open Windows and select FoxPro. On the File option select C:\ and double click on it. We get the list of directories. Then select STATMAN and click it on. This takes us to the directory STATMAN. Then type CREA BLOOD in the command window or select File option and click on New. We can then type BLOOD in the place provided for it.

We can now enter the data and after all the records are entered save the file using Ctrl + W.

The structure can be viewed as shown below using the command DISPLAY STRUCTURE.

The data can be viewed by using BROW command. The actual data obtained from the FoxPro file is shown in Appendix B. The file contains some extra fields, which we shall discuss shortly.

```
Structure for database : C:\STATMAN\BLOOD.DBF
Number of data records :      45
Date of last update     : 08/31/99
    Field   Field Name   Type        Width    Dec
    1       SNO          Numeric     2
    2       YEAR         Numeric     4
    3       MONTH        Character   3
    4       OPOS         Numeric     3
    5       ONEG         Numeric     2
    6       APOS         Numeric     3
    7       ANEG         Numeric     2
    8       BPOS         Numeric     3
    9       BNEG         Numeric     2
    10      ABPOS        Numeric     2
    11      ABNEG        Numeric     2
** Total **                          29
```

Let us answer some basic questions from this data file.

Question 1. What is the total number of donations (of all groups) in each month with positive and negative Rh values for the year 1996?

To answer this question, we fist create three new fields namely RHPOS, RHNEG and TOT to mean Rh-positive, Rh-negative and total donations respectively. To do this the command is MODI STRUCTURE. This gives the existing structure, which can be modified. We then type the new numeric fields with width of size 3. Now type the following commands to provide the sums automatically.

```
REPL ALL RHPOS WITH OPOS + APOS + BPOS + ABPOS
REPL ALL RBNEG WITH ONEG + ANEG + BNEG + ABNEG
REPL ALL TOT WITH RHPOS + RHNEG
```

(REPL is the shorcut for REPLACE command.) Now BROW and see the values in the new fields. The changed structure appears as follows:

```
Structure for database : C:\STATMAN\BLOOD.DBF
Number of data records :     45
Date of last update   : 08/28/99
     Field   Field Name   Type        Width   Dec
     1       SNO          Numeric     2
     2       YEAR         Numeric     4
     3       MONTH        Character   3
     4       OPOS         Numeric     3
     5       ONEG         Numeric     2
     6       APOS         Numeric     3
     7       ANEG         Numeric     2
     8       BPOS         Numeric     3
     9       BNEG         Numeric     2
     10      ABPOS        Numeric     2
     11      ABNEG        Numeric     2
     12      RHPOS        Numeric     3
     13      RHNEG        Numeric     3
     14      TOT          Numeric     3
** Total **                           38
```

We observe that now the three new fields are on display. The answer for the question given above is obtained by typing the following commands.

CALC SUM(TOT) FOR YEAR = 1996

This gives 2635. In order to get the monthly donations, the command is

DISPLAY FIELDS MONTH, YEAR, RHPOS, RHNEG, TOT FOR YEAR = 1996

We get the following output as a table with the field names as headings:

MONTH	YEAR	RHPOS	RHNEG	TOT
JAN	1996	165	19	184
FEB	1996	268	28	296
MAR	1996	226	20	246
APR	1996	206	24	230
MAY	1996	293	32	325
JUN	1996	134	13	147
JUL	1996	224	24	248
AUG	1996	132	16	148
SEP	1996	194	18	212
OCT	1996	232	13	245
NOV	1996	197	18	215
DEC	1996	127	12	139

Question 2. What is the average number of donations for the year 1995?

The command is CALC AVG(TOT) FOR YEAR = 1995 and we get 237.27

Question 3. What is the average number of donations in August?

The command is CALC AVG(TOT) FOR MONTH = 'AUG'. This gives the average of August for all the three years 1994, 1995 and 1996, and we get 175.0. We have to note that AUG shall be typed using single quotes since it is a character field.

Question 4. In which month there was maximum number of donations?

We can use the command CALC MAX (TOT) and we get 353 but it does not say for which month this maximum has occurred. Now type the following commands one after another. After each command we have to press the Enter key.

    SORT ON TOT TO TT
    USE TT
    GO BOTTOM
    ? MONTH, YEAR, TOT

TT is the name given to temporary file. This gives the result in three separate items as

    JUNE 1994 353

It means that, the maximum has occurred in June 1994 and the total number of donations is 353. Here we have first sorted the file in the increasing order and written it as new file called TT to mean temporary file. Then we have used TT since we have been earlier using the file BLOOD. Since the file is in the increasing order, the value of the field TOT in the last record must be the maximum. We have located this by going to the bottom of the file using GO BOTT. The last command in the list given above starts with a question mark '?' which is used to print the results on the screen. Since we need the month, year and the total, we have used the above commands all in one line. We could have alternatively used them in three separate lines like

    ? MONTH
    ? YEAR
    ? TOT

We can similarly find the minimum with the following commands

    GO TOP
    ? MONTH YEAR TOT

Since sorting is over and the results have been obtained, we may delete the file TT. To do this, the command is DELE FILE TT.DBF. We then get the message

FILE HAS BEEN DELETED

Question 5. What is the overall percentage of Rh-negative donations in the total donations?

The commands to be typed for this query are

USE BLOOD
CALC SUM (TOT)

This gives 9016.

CALC SUM (RHNEG)

This gives 759.

? 759 * 100/9016

This gives 8.42.

Consider another command
CALC AVG (RHNEG * 100/TOT)

This gives 8.08 as the average percentage of Rh-negative donations. We have to notice the difference between these two results. The first one of 8.42% is the percentage obtained over the total donations shown in the 45 records. The second one of 8.08 is based on the individual (monthly) percentages calculated for each record and then the average is found for these percentages. The percentages are not visible even with the BROWSE command. This is because we have never created a field and stored the data in it.

Suppose we wish to store the individual percentage values of RHNEG in the total donations for each record. Then we have to type MODI STRU and create a new numeric field say PERCENT with the field width of 5 and 2 decimal spaces. Then we can use the command

REPL ALL PERCENT WITH (RHNEG/TOT) * 100

Now all the 45 records will get the percentage. We may now find the average by using the command CALC AVG (PERCENT), and we get 8.08%.

3.4 FOXPRO FILE FOR FOOD GRAINS DATA—A CASE IN ECONOMICS

As a part of a major study, *Food Grain Inventory System in India,* Sarma (1998) has examined the data on production, procurement and offtake of food grains in India over several years. The objective is to forecast the future values by examining the current trends shown by the available data. Table 3.2 gives the actual data.

Table 3.2 Production, Procurement and Offtake of Food Grains (in Million Tonnes)

Year	Production	Procurement	Offtake
1970–71	108.42	8.794	7.816
1971–72	105.17	8.311	11.396
1972–73	97.03	7.553	11.414
1973–74	104.67	6.249	10.790
1974–75	99.83	8.152	11.253
1975–76	121.03	13.261	9.174
1976–77	111.17	9.832	11.729
1977–78	126.41	10.365	10.183
1978–79	131.90	14.469	11.663
1979–80	109.70	9.823	14.993
1980–81	129.59	12.317	13.014
1981–82	133.30	15.275	14.768
1982–83	129.52	15.518	16.206
1983–84	152.37	17.076	13.336
1984–85	145.54	20.431	16.441
1985–86	150.44	20.511	19.804
1986–87	144.07	17.051	22.582
1987–88	138.41	13.367	19.980
1988–89	172.18	16.719	16.300

Let us first create a data file named FOOD and save it in the directory C:\STATMAN with the following structure. The year is taken as character instead of numeric because the data contains a non-numeric character '-'.

```
Structure for database  : C:\STATMAND\FOOD.DBF
Number of data records  :      19
Date of last update     : 09/01/99
   Field Field Name  Type        Width   Dec   Index
     1   SNO         Numeric       2
     2   YEAR        Character     7
     3   PRODUCE     Numeric       6      2
     4   PROCURE     Numeric       6      3
     5   OFFTAKE     Numeric       6      3
** Total **                       28
```

Forecasting is a complicated statistical problem. Simple FoxPro commands will not directly do this job. We have to write a PROGRAM in FoxPro to get the results. We will see in Chapter 11 that a built-in facility is available in MS-Excel for this type of work.

Certain basic questions can however be answered from this FoxPro file directly. For instance, what is the average ratio of procurement to production? To answer this question, the command is

CALC AVG (PROCURE/PRODUCE)

This gives 0.10. It means that on an average, 10% of production goes to procurement.

3.5 FOXPRO FILE FOR PLANT GROWTH PROBLEM— A CASE OF EXPERIMENTAL DATA

Botanists have conducted several experiments to demonstrate that plant seeds when exposed to ultrasound treatment (UST) can have a positive effect on the growth of the plant. The duration for which seeds are exposed to UST is an important factor in such experiments.

Suresh (1999) has conducted some trials on Vigna Radiata-L (green gram) cultivars in order to examine the effect of UST of seeds on plant growth. The experimental seeds have been selected from three cultivars (varieties) coded as V-407, V-410 and V-450. From each variety the seeds are exposed to UST at varying intervals as 15-minutes, 30-minutes and 60-minutes. Some seeds without UST are also used as 'Control' to mean 0-minutes of exposure.

Table 3.3 Shoot Length of Cultivars at Different Treatments

	Varieties of seeds		
	V-407	V-410	V-450
	25-DAS		
CONTROL	15,16,15,14,15	17,16,17,16,17	15,16,14,15,14
15-MTS	19,17,16,17,19	20,21,19,20,21	18,17,16,17,18
30-MTS	20,21,19,20,21	23,22,24,23,24	20,21,19,21,22
60-MTS	15,24,23,25,24	27,28,29,30,32	26,27,28,24,23
	35-DAS		
CONTROL	20,21,20,22,21	25,24,26,24,25	20,21,20,21,22
15-MTS	28,27,29,28,27	30,32,31,33,32	29,30,29,31,29
30-MTS	36,35,34,35,36	40,41,42,41,42	38,37,36,39,38
60-MTS	44,43,44,45,43	51,52,54,55,56	46,45,44,45,46
	45-DAS		
CONTROL	29,30,29,31,30	38,35,36,37,39	28,29,30,31,32
15-MTS	40,41,42,41,40	43,44,45,46,45	40,41,42,41,40
30-MTS	46,45,47,46,45	50,51,52,53,50	48,49,47,43,45
60-MTS	54,55,56,55,56	59,60,61,59,58	54,55,56,57,54
	55-DAS		
CONTROL	40,41,40,42,41	48,46,47,49,48	46,47,48,45,44
15-MTS	50,51,52,50,51	54,55,56,57,58	54,53,55,56,54
30-MTS	56,57,58,56,59	59,60,61,60,62	58,57,56,55,57
60-MTS	60,59,61,62,61	63,64,65,66,67	64,63,62,61,63

For each variety, at each treatment (duration of UST), five replications were made using a randomized block design. Several parameters have been studied but 'shoot length' is one parameter used in this illustration. The shoot length has been observed at different *days after sowing* (DAS) namely 25-DAS, 35-DAS, 45-DAS and 55-DAS. The data from this experiment are as given in Table 3.3.

The problem is to compute the *average shoot length* and its *standard error* at each growth stage. It is also required to perform a comparison of the effect of UST on the shoot length and verify whether it varies interaction between the cultivars and the treatment level.

We first create a FoxPro file for the data given above. Let us denote the levels of treatments as 1, 2, 3 and 4 to denote Control, 15-MTS, 30-MTS and 60-MTS of UST respectively. We also denote the varieties by 1, 2 and 3 receptively.

Let us create a structure with 4 response variables SL-25, SL-35, SL-45 and SL-55 to designate the four growth stages. The experimental factors are treatment and variety.

Let us call this file as PLANT and keep it in the directory STATMAN. The structure for this file is created as shown below. The complete data with 60 records is given in Appendix C.

```
Structure for database  : C:\STATMAN\PLANT.DBF
Number of data records  :       60
Date of last update     : 09/25/99
    Field   Field Name   Type        Width    Dec
        1   SNO          Numeric     2
        2   VARIETY      Numeric     1
        3   TREATMENT    Numeric     1
        4   SL25         Numeric     2
        5   SL35         Numeric     2
        6   SL45         Numeric     2
        7   SL55         Numeric     2
** Total **                         13
```

Now let us perform some basic statistical calculations from this file. First open FOX and USE PLANT. The following commands would give the required results. The mean and standard error of SL-25 for the first variety and first treatment can be obtained as follows:

CALC AVG(SL25) FOR VARIETY = 1 AND TREATMENT = 1

This gives 15.00.

CALC STD(SL25) FOR VARIETY = 1 AND TREATMENT = 1

This gives 0.63.

? 0.63/SQRT(5)

This gives the standard error as 0.28.

> In FoxPro, the command used at a previous step can be viewed at a latter step using the edit key. FoxPro will store all the previous commands so that we can select one with the help of this edit key. This facility is known as *DOSKEY*.

Again for TREATMENT = 2 and VARIETY = 1 we can repeat the above steps with suitable modification as given below:

CALC AVG(SL25) FOR VARIETY = 1 AND TREATMENT = 2

This gives 17.60.

CALC STD(SL25) FOR VARIETY = 1 AND TREATMENT = 2

This gives 1.20.

? 2.24/SQRT(5)

This gives the standard error as 0.54.

We can repeat the above steps for TREATMENT = 2 or 3 or 4 as the case may be and get the remaining results. It is however difficult to repeatedly type the commands over a large number of variables and conditions.

The methods discussed so far could give an impression that we are using FoxPro like a calculator, which requires human intervention at every step. The only benefit is that the data in FoxPro is available as a file and we need not enter the values every time we operate a command. It would be convenient to put together all the commands as a *program* so that it can be *run* on any FoxPro file any number of times. (Some simple ideas to create a program are given in Chapter 10.)

REFERENCES

1. Lavanya, S. (1999): Project study on nutrition, Department of Home Science, S.V. University, Tirupati.
2. Richard A. Johnson (1995): *Miller and Freund's Probability and Statistics for Engineers*, 5th ed., Prentice-Hall of India, New Delhi.
3. Sarma, K.V.S. (1998): *Food Grain Inventory System in India*, project report on 'operations research for inventory control—the gap between theory and practice', Department of Statistics, S.V. University, Tirupati.
4. Suresh (1999): 'The effect of ultrasound treatment of seeds on the shoot length of plants', part of the doctoral thesis, Department of Botany, S.V. University, Tirupati.

5. Taxali, R.K. (1996): *FoxPro 2.5 Made Simple for DOS and Windows*, BPB Publications, New Delhi.
6. Vijaya Kumar, K. et al. (1997): Project report on blood bank statistics, Department of Statistics, S.V. University, Tirupati.

DO IT YOURSELF

3.1 Create a data file to enter the name and telephone number of your friends.

3.2 Create a data file to enter the daily maximum and minimum temperature for one month in your city/town using the data from a local newspaper.

3.3 Find the frequency table for the variable BMI in the file TRIBAL.DBF with classes as

14–16, 17–19, 20–22, 23–25, 26–28, 27–29

3.4 Calculate the count for below average and equal to or above average BMI.

3.5 Make a cross-tabulation of Sugali, Yanadi and Yerukala with BMI equal to or above average and below average.

3.6 Calculate the average of percentage BMI to EBAL. (*Hint:* Use the command CALC AVG (BMI*100/EBAL).)

3.7 Delete all records of males in TRIBAL.DBF file. (*Hint:* DELE ALL FOR SEX = 2.)

3.8 Recall all the deleted records in Question 3.7.

3.9 Type F1 or Help in the command window of FoxPro and examine the details therein.

Windows and MS-Office for Research Studies

The invention of Windows operating system by the Microsoft company has brought out a remarkable change in the use of personal computers. Those who were earlier working in DOS environment find it a loving job to work with Windows.

The researcher needs certain facilities like creating text documents, calculations, tables, graphs, slides for presentation, etc., as a part of the research. All these jobs could be done with the help of a package called MS-Office, which works in the Windows environment. Certain statistical packages like SPSS and SAS work in Windows operating system. So it is useful to understand how Windows and MS-Office can help the researcher.

4.1 ABOUT WINDOWS 95/98

Microsoft Windows (MS-Windows) is a single-user operating system for personal computer. It provides a *graphic user interface* (GUI) for carrying out common routines like creating directories, copying or renaming files, etc.

When the system is booted, it normally goes to Windows. If not, it means that the path is not set to go to Windows directly. This effect can be obtained by changing the parameters in DOS system files. In that case, the system stops at the DOS prompt C:\ and we have to type the command WIN. This will open Windows.

There is an advantage in setting this type of booting sequence, namely DOS and then Windows. When we have to work only in DOS we can directly start working. For instance FoxPro 2.5 of DOS version, which we have discussed in Chapter 3 can be opened without opening Windows. Suppose we wish to record a few sentences in the form of a note, we can do it either in DOS Editor or WordStar. If the researcher is a user of programming languages like COBOL, Pascal, C, BASIC, etc., these can be worked directly in DOS and there is no need to open Windows.

On the other hand, suppose the system goes to Windows directly. We can then select the MS-DOS program from the Program files and use the DOS commands. If we use the combination Alt + Enter, we get the DOS screen with Windows screen as background. But we should remember that we are keeping open a large number of Windows files just for working in a software which does not require Windows. So, it is advantageous to open in DOS and then switch on to Windows only when required. After typing WIN, Windows will open with the screen as shown in Figure 4.1, which is also known as *desktop*.

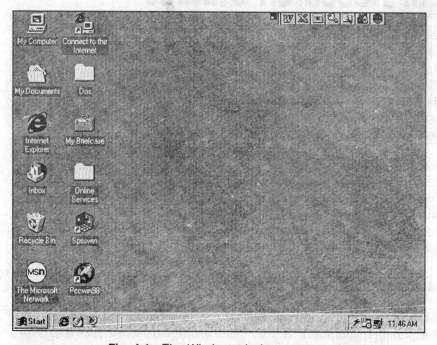

Fig. 4.1 The Windows desktop screen.

By clicking on any of these items on the desktop, the corresponding application will be activated. All those software applications, which we wish to directly open from the desktop can be kept as icons on this screen. For instance, we can have an icon for anti-virus, an icon for MS-DOS and so on. We also find another item on the desktop, called Recycle Bin, which stores all the deleted files. At the bottom of the screen we find a *task bar* on which the Start button appears. By a click on Start we get the menu indicating different items. The major items on the Start button are shown in Figure 4.2. Black arrows (▸) against items in the menu indicate that there are some options available on that particular items. By selecting such an item with the mouse or with the arrow keys, we get another screen to the right of it, in which various executable files will appear. Among them if we click on MS-DOS, the control goes to the normal DOS screen and all the DOS commands will work.

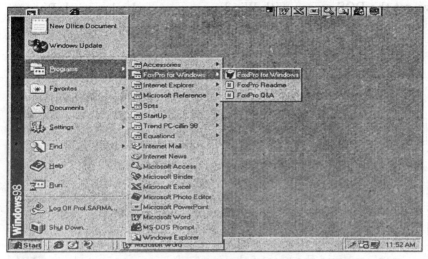

Fig. 4.2 Windows Startup menu.

The following are some salient features of Windows, which have made it more interesting than the DOS.

- The commands are visible on the screen.
- The screen in Windows provides a graphic display of information. Because of this facility, graphs and diagrams can be handled very efficiently.
- We can select any command with the help of the mouse. By moving it on the mouse-pad, a cursor (a pointer usually in the shape of an arrow) moves on the screen.
- The mouse usually contains three buttons. The left click of the button is used to activate a command. For instance, by clicking with left button on the Start icon, we get the corresponding screen on it. The right button of the mouse is useful to view the properties corresponding to an item. Not all icons will respond with the right click. If there is anything to be conveyed on a particular icon or message, the right click will respond. The middle button has special applications like controlling the flow of information on the screen. This is called *scrolling* and very useful in reading a document in MS-Word.
- The commands namely COPY, MOVE, RENAME or DELETE can be applied very efficiently without taking care of the *syntax*. (In DOS, we have to type the command without a single mistake!). We can also *create* a *directory* with a simple graphic help on the screen.
- Whenever a doubt arises about a command, an on-line help can be obtained by clicking on Start ▸ Help.

- The most interesting facility in Windows operating system is that when a file is deleted form a directory it will ask for a confirmation and if we click 'Yes' it will be deleted. The deleted file moves to a separate directory in the hard disk called the Recycle Bin. We can see the list of deleted files by clicking on the Recycle Bin with mouse and then clicking on the Explore option. Suppose we wish to retain the file, which was already deleted, we first click on that file and then click Restore. The file goes back to the location from where it was deleted.
- Much of the user-software is now available in Windows environment only. As a result, we can *navigate* from one application to another. For instance, while working in a statistical software like SPSS, we can copy the output from it, open another software like MS-Word and paste this output at the desired place. During this time, the SPSS will still be running and we can continue the rest of the work after saving and closing MS-Word.
- This is the age of information technology and communications play a dominant role in all sectors like education, research, training, business, health and industry. MS-Windows provides a software to connect the PC with various users all over the world. This is available from the desktop item called Internet Explorer.

In the following section we will look at file handling using a program called Windows Explorer.

4.2 THE WINDOWS EXPLORER

A powerful tool in Windows is the Windows Explorer. It is a graphic and tree-like display of various directories, sub-directory and files. In Windows terminology, a directory is called a Folder and we can have many sub-folders within a Folder. The following are some operations with the Explorer. The Explorer screen looks as shown in Figure 4.3.

In order to open Explorer, we can use the keys

Start ▸ Programs ▸ Windows Explorer

We can alternatively use the toolbar corresponding to Explorer. Some of the facilities under Explorer are as under:

My computer

We can click on this desktop item to view the different software and their locations. By clicking on any item, another screen with details

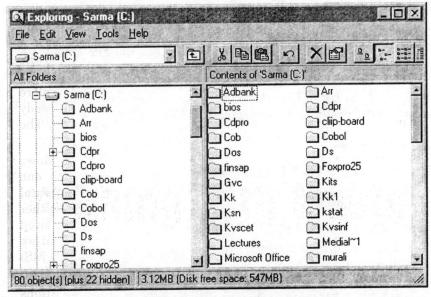

Fig. 4.3 Windows Explorer screen.

appears. We can give a right click on any item and look at its properties. The MYCOMPUTER screen appears as in Figure 4.4.

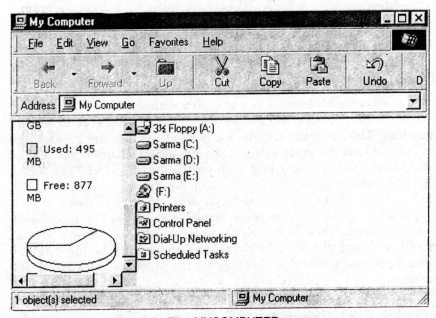

Fig. 4.4 The MYCOMPUTER screen.

Creating a folder

Let us create Folder by name KVS with the help of Windows Explorer. To do this, choose Windows Explorer and select the File option with the help of the mouse. Then select the option New Folder and click on it. A new folder without a name appears in the bottom-most position of the list of files. There, we have to type the name KVS at the blinking position. This creates the folder KVS.

Copying a file

Consider the file TRIBAL.DBF, which is available in the directory STATMAN. Let us copy it to the folder KVS. To do this with explorer, proceed as follows.

- Select the folder STATMAN and click on it
- The list of files appears, from which we select TRIBAL.DBF and right click on it
- We get a pull down menu with items like CUT, COPY, PASTE, MOVE, RENAME, etc.
- Select Copy and click
- Now select the directory KVS and right click there
- Choose Paste option and click on it
- The file TRIBAL.DBF gets copied and appears in the KVS directory.

As an alternative we can click on TRIBAL, drag it with mouse on to the folder KVS and leave the mouse. The file gets copied.

The FIND command

It is an easy job to locate a file or folder with the help of Explorer. We can verify whether the file we need exists at all, if so in which directory. The command for this job is FIND. It is found by clicking on Tools option in the top menu of the Explorer. We have to type the name of the file or folder and click on the option called Find Now. The drive where the search has to be made can also be specified. All the locations where the desired file is available will be listed in the output window. We can thus check whether the desired file is available or not. Windows then starts searching for it. All the locations where the file is found are listed in a sequence along with the file details. This is extremely useful in checking for the presence of files or their location.

Creating a desktop item

Once the file or directory is found, the right click on that gives a

small menu in which we find an item named Send To. If we click on it, we find the option Desktop As A Short Cut. If we click on it, a small icon on the desktop will be created so that whenever Windows is opened, we can simply double-click (two left clicks without delay) on this icon and the corresponding application works. The following are the steps to create a desktop item:
- Open Explorer
- Select Tools
- Click on Files or Folders
- Type the name of the file or folder to find
- If found, right-click on the name in the list
- Select Send To
- Select Desktop As A Short Cut.

Operations with floppy disk

A file or a group of files or a directory can be copied from hard disk to a floppy disk with the help of Windows Explorer. Suppose we wish to copy the file TRIBAL.DBF to the floppy. The following are the steps to do this job.
- Open Windows Explorer and select the directory STATMAN
- Select the file TRIBAL.DBF and right click on it
- Select the Send To option
- Click on 3.5 inches floppy diskette
- The file gets copied to the floppy disk (designated by A:\) and an animation of this movement appears on the screen.

It is worth remembering at this stage that the usual DOS command for this job is

COPY C:\STATMAN\TRIBAL.DBF A:\

Suppose we wish to copy the entire directory STATMAN to A: with all its files and subdirectories, if any To do this we proceed as follows:
- Choose the folder STATMAN and right click on it
- Click on the Send To option
- A new directory with name STATMAN will be created in A:\ and the contents of this folder will be copied to A:\.

In a similar way, the contents of a floppy disk can be copied on to a specific directory in the hard disk. Suppose we have a directory named HOSTELS in A:\ with a few files in it. We wish to copy the

same to the hard disk. To do this we proceed as follows:
- Open Explorer and select C:\
- Now select A:\
- The directory HOSTELS will appear in the list
- Right click on it and select Copy option
- Select C:\ and right click on it and select Paste
- The directory HOSTELS will be copied to C:\.

We can also use the Explorer to examine the properties of a file or a directory or the hard disk itself. Select C:\ and right click on it. A pie diagram appears showing the amount of space used up and the space still free on the disk can be seen (see Figure 4.4).

File properties

For any directory or file we can see the file properties by a right click on the corresponding file or folder. We get several details among which the attributes of the file are indicated. The file name, its type, date of creation, date of modification, etc., are displayed.

We can also check up whether the file is *Archieve* or *Read-only*. A file, which is Archieve, can be used for both Read and Write operations while a Read-only file cannot be modified. The screen for file properties is shown in Figure 4.5 with reference to the file TRIBAL. Another important use of file properties is to see the size of the file

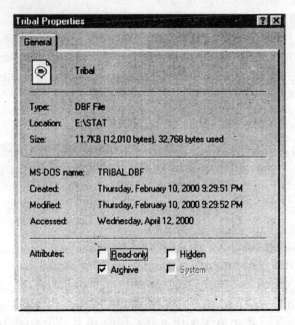

Fig. 4.5 File properties.

in terms of the bytes. This helps in comparing the space available in the floppy disk with the file we wish to copy. However, the Copy option will automatically does it and warns that there is insufficient space, whenever the size of the file to be copied is larger than the space available on the target disk.

Compressing a file

Suppose a file in the hard disk has occupied a space of more than 1.44 MB. It is not possible to copy this file to A:\ due to space constraint. We can use special software called Winzip. This would compress the file so that it uses less number of bytes than the original file. When required, this file can be *unzipped* or *extracted* into the original version with the help of Winzip.

Deleting files

We can delete a file by selecting the file and then giving a right click. The property screen appears in which one of the item is Delete. If we click on this item, a message appears asking whether the file is to be sent to Recycle Bin. If we click Yes the file will be sent to the Recycle Bin. If we wish to permanently delete, we can use the key combination Shift + Del. We can apply the same procedure for multiple files by blocking them and deleting. To block a group of files we place the mouse cursor at one file and then use Shift + Arrow key and move over all the files to be blocked. All the files of a directory can be selected at a time by using the command Ctrl + A or by choosing the Select All option in Edit menu.

Thus we can very efficiently use Explorer for file handling.

4.3 OPENING FOXPRO IN WINDOWS

The entire discussion made in Chapter 3 regarding FoxPro data files and executing programs can be done by using a software called FoxPro for Windows. It does not work in DOS mode.

We can also use the software called Visual FoxPro, if it is available. The opening FoxPro screen for windows appears as shown in Figure 4.6. In order to select the file, we first double-click on C:\. This shows all the directories. Then the required directory can be clicked. This will open the directory with the list of files shown in the left half of the window screen. The required file can be selected by a double-click on it or by one click and then pressing Enter key. To see the contents of this file we have to either type BROW in the command windows or click on the top menu items in the order,

Database ▸ Brow

Fig. 4.6 FoxPro opening screen in Windows.

Once the file is browsed, the usual command window appears and we can type all the commands, which we were using with FoxPro in DOS mode.

Operations about file-editing, like APPEND, DELETE, etc., can be carried out with the help of mouse from the menu that appears on the top of the screen.

A FoxPro program like DSTAT, given in Chapter 10, can be run in this software by selecting the option RUN and then clicking on DO. The list of PRG files will be on display from which we can click on the desired file. FoxPro will automatically run it.

4.4 ABOUT MS-OFFICE

MS-Office is an integrated package with user-friendly commands and graphic utilities. It works in Windows environment (operating system) like Windows 95 or Windows 98. It is available in most of the personal computers. It contains four major components: MS-Word, MS-Excel, MS-PowerPoint and MS-Access.

We shall look at some details of these facilities and their use for the researcher.

MS-Word

It is a word processor useful for preparing text, tables, equations, symbolic representation and even pictures. A research proposal can be prepared in an elegant way by using Word. The job of typing the text in this software is easy with its stylish fonts and sizes. Different models (templates) are available for specific jobs like making a business report, a resume, a legal document and so on. When a new file is opened, Word has an option to choose the template that suits our requirement. If nothing is chosen, it goes to Normal template, which is used to prepare a general document.

MS-Excel

It is a big worksheet used for data entry and to perform calculations. It is also used to create tables, graphs and also to perform some advanced statistical calculations. For a user, Excel is like a wonder world since it makes calculations very simple. In addition to this, a part of the work done in Excel can be copied and pasted into a Word document.

MS-PowerPoint

It is a package used to create slides and transparencies involving graphics, text and drawings with high quality. The researcher often needs a quick method of presenting the findings without loosing valuable information. One can also create animation effects in a presentation. For instance, if an economist wishes to show the impact of a pricing policy on reducing the cost of living index, then PowerPoint is the best tool. New product demonstrations, financial status of companies are usually presented with the help of PowerPoint.

MS-Access

It is a powerful tool for maintaining databases. It contains a provision to prepare a database in a way easier than that of FoxPro or Excel. Once a new database is created, several queries can be answered from the data and we can create forms, tables and reports. We can also interact with other databases using a network.

Thus MS-Office is an integrated package to maintain a database (with Access), to create graphs and calculations (with Excel), to type a report (with Word) and to create high quality presentations (with PowerPoint). Naturally, it should help every user of a PC without looking for separate software for each of these jobs.

Office toolbars

This is a very useful item for the users of MS-Office. While installing MS-Office, this toolbar is automatically created in the Startup process. The toolbars (also known as shortcut bars) which appear on the top-right corner of the desktop screen (see Figure 4.1) can be used to activate Word, Excel, PowerPoint or Explorer by a single mouse-click. We can also customise the toolbars by a right click on the left most position of the toolbars. By selecting the item Always On Top we can keep the toolbars always at the top of the screen irrespective of the software in which we are working. If we include Program Files in the toolbars, we can get any executable file directly from this toolbar.

Suppose we are using Word and somewhere we wish to see a data file or graph available in MS-Excel. We need not close Word but can open Excel by directly clicking on the corresponding toolbar. During this time Word can be minimized by clicking on the '–' button that appears on the top right corner of the screen. Now we can select the graph or data and copy it. The portion can be pasted in the Word document very easily at the required position. We can use Ctrl + C for *copy* and Ctrl + V for *paste* operations. We can keep other software packages like MS-DOS, FoxPro, etc., in the toolbars so that they can be readily accessed.

4.5 ABOUT MS-WORD

WORD can be opened with the following toolbar, which appears on the top of the screen. By clicking on the button **W**, the Word screen appears with its logo as either Word 97 or Word 6.0 depending on the version available. We can open Word in the following sequence with mouse:

      Start ▸ Programs ▸ MS-Office ▸ MS-Word

After a few seconds, the screen appears with an empty document ready to type text. The top-left portion of the screen displays the name of the document as Microsoft word document1. If we open an existing document, its name will be displayed in place of document1. Several icons appear on the screen to help the user on different aspects of preparing the document. Some important aspects are briefly discussed in the following paragraphs.

The main menu

The main menu of Word contains the following items on the top row of the screen, as in Fig. 4.7. By clicking on any items on this menu, the corresponding sub-menu appears.

| File | Edit | View | Insert | Format | Tools | Table | Window | Help |

Fig. 4.7 The main menu of MS-Word.

For instance if we click on File or press Alt + F (both keys at a time) on the keyboard, the pull-down screen appears as shown in Figure 4.8.

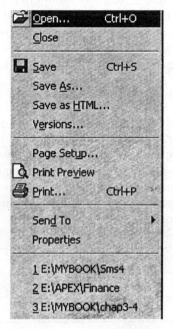

Fig. 4.8 Word screen for file handling operations.

Various modules supported by each object appear in the pull-down menu. For instance, the item View shows all options regarding the appearance of both the screen and the text on it. If we select the Page Layout option, the text appears with solid divider between pages. We can also have a Ruler option so that the margins can be maintained according to our choice. The Toolbar option has another sub-screen. It shows the items such as Auto-text, Standard, Formatting, Drawing, etc. If we click any item, it will be indicated by a ☑ mark on the left. By default the toolbars named as Standard and Formatting appear on the top of the screen. They help in opening a file, saving a file, printing a file, creating tables, etc. One important icon that comes with this toolbar is the Undo icon indicated by a reverse arrow. This is of great help in editing a text. It takes us back to a previous step. For instance, we might have converted a few sentences into Italics but wish to go back to the old pattern. It is enough to click on the undo icon. We can Undo several earlier operations one after another. Thus the toolbars are very important in working with MS-Office.

Fonts

A *font* is the style of a letter. Various types of fonts are available in Word to type a text. We can choose very specialized Font types and sizes to display the matter effectively. Fonts like *Draft, Arial, Courier New* and *Times New Roman* are commonly used for preparing text. We can change the font from one type to another by selecting the text and clicking on the new font to be used. The font that is in use is always shown in the message box of the menu.

Tables

A *table* can be created in Word by clicking on the Table option and specifying the number of rows and columns needed. The menu asks for the width of the columns and we can specify the dimension like 3 cm, 5 cm, etc. This gives a table with the desired column width. If we do not specify the column width, it goes as *Auto* and the table stretches to the full width of the page. Following is a table with 3 columns and 2 rows under Auto Width:

We can reduce the width or height of a row or column by clicking mouse on vertical or horizontal line and dragging it towards the required position. In the table menu, there is a provision for distributing the columns or rows evenly. A right click on it will distribute all columns equally without exceeding the boundaries set.

Now let us redesign the above table with a column width of 5 centimetres. The table is initially aligned to the left. Now click on Table option and click on Select Table. Then by pressing Ctrl + E, the table will be centred and looks as follows. We can also use the icon for centering.

We can change the style of the table by selecting Table Autoformat option. A complete table or a part of it can be copied and pasted at different places of the document. This is useful when the research report requires a large number of identical tables to be created with different data. We can copy an existing table and fill with a new data by simply over-typing on the old data.

Making bibliography and referene lists

We can make use of tables in Word to create a list of references or a bibliography of various books and journals. A reference list usually contains the name of Author, the Title, the Publisher, the Years of Publication and the Key Words. We can create a Table and enter the details one after the other. We can also have a column for serial number (SNO). Now we can use the option Sort in the Table. It would sort the details in the alphabetical order of the data in the field chosen by us. For instance we can sort on AUTHOR, YEAR or on both. We can also sort on more than one field. As a result, the preparation of bibliography is an easy job.

Suppose we wish to locate all the references in which the Key word is Sampling. We can click on the column headed Key Word and block the cells. Then use the Find command of Edit menu. Type Sampling against the word to be searched. Word would highlight all the rows in which the key word Sampling is available. This is very useful for preparing research references.

We can also use this facility to prepare a *mailing list*. We can make a telephone directory with name and telephone numbers and then use sort to arrange the names in order. We can also search for a name by using the Find command.

Mail Merge

Suppose we wish to contact several persons/companies/libraries with a common message or letter. It is enough to type only one letter with provision to change the address in each letter. All addresses can be kept in a different file and the Word option called Mail Merge can be used to prepare all letters with corresponding addresses selected from the file. This is found in

Tools ▸ Mail Merge

Auto Text and Auto Correct

Certain standard phrases can be set as Auto Text so that whenever we have to use them we can insert the same without typing. For instance, we may define a closing phrase like 'yours sincerely' and whenever we wish to close a letter, we simply insert it from the list of closing options of the Auto Text.

We can use the Auto Correct facility to type a frequently occurring word or phrase without actually typing it. We can use a code with two or three letters to mean the actual word/phrase. For instance, the phrase 'S.V. University College' might be needed at many places in a document. We can define a shortcut for this in the Auto Text. To do this the sequence of mouse clicks is

Insert ▸ Auto Text ▸ Auto Correct

Under *Replace* option, we may type SVU as the shortcut and under the *With* option type S.V. University College. Then click on the *Add* button and close the Window. Here, after whenever we type SVU it will be automatically typed as 'S.V. University College'.

Text editing

How do we highlight a particular part of a text? We can make it 'Bold' or put it in 'Italics' or 'Underline' the selected text. Suppose, we wish to highlight the sentence "the difference is significant at $p = 0.01$ level". To do this we have to place the mouse at the beginning of the sentence, give left-click and drag it till the end. This is called blocking. Now click on **B** to make bold, *I* to put in italics and U̲ underline.

A part of the text, which was typed in lower cases (or small letters), can be converted into upper cases (or capital letters) by clicking 'aA' button. We can align the text to the right, centre or to the left by selecting the corresponding icon.

Any selected text or the entries in a table can also be aligned left or right or centre by blocking the text and selecting the required option. Normal text is typed with JUSTIFIED option because it will justify both left and right margins. We can also use keyboard as follows:

Ctrl + R   for right align
Ctrl + L   for left align
Ctrl + E   for centre align

A very attractive feature of Word is giving bullets as a mark of special importance to a text. On the Format we get a menu in which a provision is available for bullets. We can click on it and select the type of bullet. We can also create a menu icon for this purpose by customizing the toolbars. This aspect is very useful while making the final report or while making a presentation of transparencies or slides.

Symbols, subscripts and equations

The concern of a user in sciences is that of typing different symbols and equations. Word provides a variety of symbols. By clicking on Insert option in the main menu we can go to the item labelled as Symbols and click on it. We then get a window with all symbols. We can use Greek letters like α, β, γ or output special symbols for chemistry, botany and other applications. For those symbols, which occur frequently, we can define key combination to type these symbols and reduce the use of mouse. To do this, the following operations are used:

- Select a symbol like α with mouse from the insert menu.
- Define a short cut using one of the function keys F1, F2, etc., along with a letter to remember the key. For instance we may use F10 + a to define α and F10 + b to define β.
- After defining a short cut key, click on the ASSIGN button and close.

Typing subscripts, superscripts and complicated mathematical equation is a common need for a user in sciences. We can use the toolbar for this purpose as follows.

- Select View ▸ Toolbars ▸ Customize
- Select Command ▸ Format
- Examine the list of icons and choose the icon x^2 that gives superscript.
- Drag it with left click on the mouse and drop the icon in the toolbars. Now, this icon can be used to type a superscript. We can use Ctrl+'+' to get the superscript and Ctrl+'=' to get the subscript.
- We use a special editor called Microsoft Equation Editor denoted by the icon $\sqrt{\alpha}$ to create complicated mathematical equations (see Figure 4.9). Customizing the toolbars as done earlier, we can get this on the list of icons on the menu bar. The MS-Equation Editor is shown in Figure 4.10.

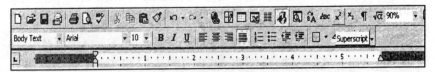

Fig. 4.9 Word tool bars.

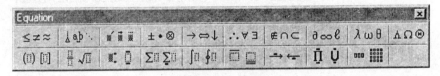

Fig. 4.10 The MS-Equation Editor.

This Editor is extremely useful to preparing scientific equations in documents.

The following is an equation typed with the equation editor:

$$\frac{\int_a^b f(x)\,dx}{\int_a^b g(x)\,dx} = \beta^{1-\alpha}$$

Saving a file

When Word is opened, it shows an empty document named as document 1. We can create a document by typing a few lines of text. We should save the file with a suitable name. Pressing Ctrl + S does this or by clicking the sequence of buttons File ▸ Save. Since no name is given to the file, it will be saved with the first line of the text as the name of the file. Further this file will be saved in the folder 'My Documents' because we did not specify the directory in which the file should be saved.

A better method is to create a folder using Windows Explorer and save the files in it. We already have a folder KVS. Let us name the current file as STAT-1. By clicking the option Save As in the File menu, the screen appears with a dialogue box to type the file name. There we have to type C:\KVS\STAT-1. If we are already working in C:\KVS we can simply type the file name as STAT-1 and we need not give the folder name.

Suppose, there are many files created in C:\ and we wish to open one of them. The option Open File will display the details of the directories (and we select among them the KVS Folder). By opening it with Open button, we get the list of files in it. We choose the required file and click on it once. Then by a click on Open, the required file will be opened. Alternatively, we can double-click on the desired file and the file opens.

We close this chapter with the observation that MS-Office is useful package for the researcher for preparing reports. Every project on statistical analysis requires a detailed report. Word helps in preparing such reports.

We will see in the next chapter some specific uses of MS-Excel and its applications to data handling.

REFERENCES

1. Microsoft Office reference from online help.
2. Robert Cowart (1995): *Mastering Windows 95—The Windows 95 Bible*, BPB Publication, New Delhi.

DO IT YOURSELF

4.1 Open Windows and right-click on the blank screen. This gives a small menu. Click on the item Properties. You can select a different background for your screen.

4.2 A new shortcut can be created on the desktop for your normal work. Right click on screen and click on New. You will find a facility to create a new desktop item. Create such a shortcut to MS-DOS by this method.

4.3 Select Windows Explorer and create a new Folder. What happens if you Move the Folder to a different location like A:\? Do you find the same again in C:\?

4.4 In Windows Explorer, files and directories usually appear as a list. What happens, if you change the View type? Select View ▸ Large icons. The view gets changed and the files appear as icons (like Briefcases) and no details will be found.

4.5 Click on any Folder or File and then right-click. Examine its properties in detail.

4.6 Select any Folder and send it to Desktop as a shortcut.

4.7 Using Explorer, find the file Toolbars. Observe that it may appear at more than one place.

4.8 Suppose you have already created an equation using the Equation Editor. You can edit it by a right-click on it and choosing the option Edit or Open. By selecting the Text option, you can convert the Italics into normal text within the equation.

4.9 Click on *bullets and numbering*. Customize it and try to change the bullet style.

4.10 There are shortcuts already given to subscript and superscript modes of typing a text. Create a new shortcut of your own choice. (*Hint:* To do this, choose the options View ▸ Toolbars ▸ Customize ▸ Commands ▸ Insert.)

Data Handling in Excel 5

MS-Excel is a part of the MS-Office package. It contains many facilities for data handling and analysis. For a regular user of MS-Excel it is like a big worksheet in the form of a table. Any type of data can be entered in the cells of the table but calculations can be done for numeric data only.

For a user, Excel is helpful to perform several statistical routines without writing programs and also to prepare high quality statistical graphs to represent data. The chief facility in Excel is that it has relative cell address, which automatically upgrades the calculations in a file whenever a value in a related cell is changed. Suppose the price indices of a commodity have been calculated taking 1975 as the base year and the complete results including graphs have been prepared. Suppose it is desired to modify all the values using 1976 as the base year instead of 1975. It is enough if we change the base year from 1975 to 1976 in the particular cell where it was typed. This will not only give new index numbers but also changes the corresponding graph. This is one great advantage for the user while carrying out repeated operations on the worksheet.

5.1 GETTING STARTED WITH EXCEL

Excel can be opened with the help of toolbars by clicking the Excel button or through the following sequence of clicks.

Start ▸ Programs ▸ MS-Excel

After a few seconds, Excel opens with its logo and a blank worksheet appears with several rows and columns. On the top of the screen, the name of the file appears as Microsoft Excel Book1. Until we specify a different name, this file name continues to appear. Even if we save the file, it will be saved as Book1, unless we specify the name. Each Excel book usually contains three worksheets designated as Sheet1, Sheet2 and Sheet3. Sheet1 will always be active and ready for data entry. We can shift to Sheet2 or Sheet3 with the help of the mouse by clicking on the corresponding sheet. Each sheet in a workbook can be used to create one data file. If there are two separate files, we can use Sheet1 for the first file and Sheet2 for the other file. The opening screen of the Excel appears as shown in Figure 5.1.

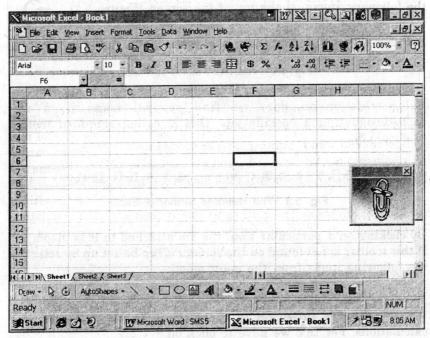

Fig. 5.1 The MS-Excel workbook.

5.2 THE EXCEL WORKSHEET

A worksheet is made up of 256 columns and a large number of rows. The columns are denoted by the letters A, B, C, ..., Z, then AA, AB, AC, AD, etc., and the last column is denoted as IV. The rows are numbered as 1, 2, 3, ..., 65536. To see this, press Ctrll + down-array key.

The worksheet is a collection of cells where each *cell* is the intersection of a row and a column identified by its column and row index. For instance, the cell belonging to column A and row 1 is denoted by A1. Similarly G10 is the cell that belongs to column G and row 10. Each column can be used to enter data that belong to one category or one vertical array of a table. A row can be used in a similar manner.

The notation A3 . . A10 or A3:A10 indicates the collection of all the cells in column A from 3rd row to 10th row. Similarly A1..D10 denotes the rectangle of cells from column A to column D in the first 10 rows. We can block a group of cells by clicking on the first cell of the desired group and dragging the mouse till the last cell of the group and leaving it there. This is very useful to copy cells from one file to another.

This way of identifying cells is essential for doing calculations on the worksheet. Once Excel is opened, the main menu contains the following items on the top of the screen (Figure 5.2).

Fig. 5.2 The main menu of Excel.

Below this menu bar, we find several icons in a row each designating a specific job. This is called *standard toolbar* (Figure 5.3).

Fig. 5.3 The standard toolbar of Excel.

By clicking on a particular icon, the job attached to it is displayed. If this toolbar is not found on the screen, it can be set up by selecting the following items from the main menu:

View ▸ Toolbars ▸ Standard ▸ Formatting

We also need a provision to type simple formulae to carryout calculations. For this we have to click on

View ▸ Formula Bar

It is interesting to note that the Office toolbars present on the top of the screen will always be available so that while working in Word it is possible to select Excel, carryout some work and return back to Word. This will be a frequent requisite to the researcher while preparing tables and graphs. If we have opened Excel while working with Word, the icons of both Word and Excel appear at the bottom of the screen in the taskbar as in Figure 5.1. By clicking on any of them the corresponding item will be activated.

5.3 DATA ENTRY ON THE WORKSHEET

Data entry in the worksheet is an easy job. We need not create a structure as we did in FoxPro. Different fields can be assigned to various columns and the field names can be typed in the first row. We call them the column headings. Both the numeric and character data can be entred in any cell. There are several types of data formats available, which can be viewed by clicking

Format ▸ cells

This gives a screen as the one shown in Figure 5.4 in which various types of defining the data are given.

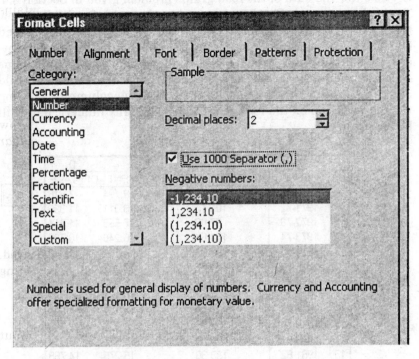

Fig. 5.4 Defining the cells in a worksheet.

Once the type of cells is defined, it is easy to enter the data without taking care of the format. For instance, if we are entering data on date, month and year we may choose the date field. If we click on Date in the menu, we get several options for entry of Date and we can select one among them. This is very useful while handling time-series data. The default setting will be General in which one can enter both numeric and non-numeric data.

When column headings (field names) are typed in the first row of a sheet, it is possible that certain names exceed the available width of the column. For instance if one field name is PROCUREMENT and the next one is SALES we find that PROCUREMENT appears as PROCUR only. The remaining part will be hidden. To see this in full, the command sequence is

Format ▸ Column ▸ Auto Fit Selection

Then the column width will be automatically adjusted to fit the heading. The row widths can also be adjusted in a similar manner, but usually not required.

5.4 DATA SHEET FOR THE FOOD GRAINS PROBLEM

Let us enter the data of the food grains problem given in Section 3.4. To do this we open Excel and straightaway enter in the first row the headings as YEAR, PRODUCTION, PROCUREMENT and OFFTAKE in columns A, B, C and D. There are 19 rows for this data. When we enter data it will be automatically right aligned as shown in the columns B, C and D of Figure 5.5. If we wish we could align the data to the centre or to the left of the column by clicking the relevant icon on the format menu bar. Usually numeric data will be right aligned.

	A	B	C	D
1	YEAR	PRODUCTION	PROCUREMENT	OFFTAKE
2	1970-71	108.42	8.794	7.816
3	1971-72	105.17	8.311	11.396
4	1972-73	97.03	7.553	11.414
5	1973-74	104.67	6.249	10.790
6	1974-75	99.83	8.152	11.253
7	1975-76	121.03	13.261	9.174
8	1976-77	111.17	9.832	11.729
9	1977-78	126.41	10.365	10.183
10	1978-79	131.90	14.469	11.663
11	1979-80	109.70	9.823	14.993
12	1980-81	129.59	12.317	13.014
13	1981-82	133.30	15.275	14.768
14	1982-83	129.52	15.518	16.206
15	1983-84	152.37	17.076	13.336
16	1984-85	145.54	20.431	16.441
17	1985-86	150.44	20.511	19.804
18	1986-87	144.07	17.051	22.582
19	1987-88	138.41	13.367	19.980
20	1988-89	172.18	16.719	16.300

Fig. 5.5 Excel worksheet for the food grains data.

Names and other character data may be left aligned for a proper presentation. In order to modify an existing alignment pattern of a column, we have to block the column and then click on the alignment icon. After entering all the data, the worksheet appears as given in Figure 5.5. We can save this file with a name by selecting the sequence

File ▸ Save As

At the cursor position in the window, we have to type the name of the file preferably with the PATH (name of the directory where to save the file). If a new directory is to be used for this purpose, it is suggestive to create a directory in Explorer or My Computer and use that name in this path.

DATA HANDLING IN EXCEL

One of the options is the file name and file type. If we do not specify the type of the file, it will be saved as Microsoft Excel workbook and the file extension will be XLS. Some beginners have a tendency to save a file without specifying its name. Then all such files will be saved in the folder My Documents as Book1, Book2, Book3 and so on. It will be very difficult to distinguish among these files without proper names. In fact it is a good practice to first select the directory into which the files should go.

The name of the file will appear as Book1. Let us save the file as C:\STATMAN\FGRAINS. The name of the file including the name of the directory can directly be typed in the place provided against file name and we type the above name. After this, we can either press Enter or click OK button. This will complete the process of saving the file. Whenever some modifications have been made to this file, we can save them by simply by pressing the combination of keys Ctrl + S.

The file is then saved as Excel workbook because the file we have used is Microsoft Excel workbook as displayed in the window against File Type. By clicking at the scroll pointer at this window, we get various options to save the file. For instance, we can save the same file as DBF 3 to mean dBASE III file by selecting the file type as dBASE III. A FoxPro file is also called a dBASE file. If we do not specify the type, it will be automatically saved as Excel file and the extension name of such a file would be '.XLS'.

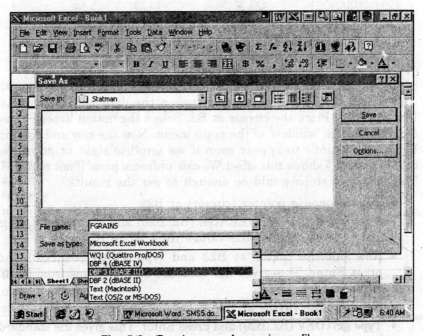

Fig. 5.6 Excel screen for saving a file.

Figure 5.6 shows the dialogue box corresponding to file saving operation.

5.5 CALCULATIONS ON THE WORKSHEET

We can perform routine calculations like sum, difference, average, standard deviation, percentage, etc., on the worksheet itself. Consider the data shown in Figure 5.7. Let us find the average of production, procurement and offtake over all the years and put it in the last row

	A	B	C	D	E
1	YEAR	PRODUCTION	PROCUREMENT	OFFTAKE	
16	1984-85	145.54	20.431	16.441	
17	1985-86	150.44	20.511	19.804	
18	1986-87	144.07	17.051	22.582	
19	1987-88	138.41	13.367	19.980	
20	1988-89	172.18	16.719	16.300	
21	RESULTS				
22	AVERAGE	126.88	12.90	13.83	
23	VARIANCE	423.27	18.79	15.54	
24	STDDEV	20.57	4.33	3.94	
25					

Fig. 5.7 Basic statistical values obtained using Excel commands.

of the table. To do this, we go to the last empty row (21st row) with a provision of one row as a gap between the data and the results. Here is a tip. Place the cursor at B2. Select the option 'freeze pans' from the option 'window' of the main menu. Now the row and column headings continue to appear even if we scroll to right or down the cells. Figure 5.7 shows this effect. We can 'unfreeze pans' if not required. The following steps would be enough to get the results:

- Put the mouse pointer (cursor) at B22
- In this cell, type @AVERAGE (B2..B20) and press Enter. This gives the average of the values from B2 to B20.
- Now put the cursor at B23 and type @VAR (B2..B20) and press Enter. This gives the variance of the values from B2 to B20.
- Now put the cursor at B23
- Type @STDEV (B2..B20) and press Enter. This gives the standard deviation of the values from B2 to B20

- In order to perform these calculations for the other two columns we need not repeat this procedure again. We can use facility called Copy.
- Block the three cells B22, B23 and B24 with the help of the mouse.
- Now copy the contents either by using Ctrl + C or by selecting the copy option from the EDIT menu.
- Select the cells C22 to D24 and press Ctrl + V or Paste the contents from the Edit menu.

This gives corresponding results for the other two columns.

As an option we can type the labels AVERAGE, VARIANCE and S.D. in the first column against the results. Suppose we wish to calculate the percentage of offtake to the procurement and wish to record these values in column E. To do this, select column E and type PERCENTAGE as the heading. The remaining steps are as follows.

- Select the cell E2
- Type the formula as +D2 * 100/C2 and Enter
- This gives the percentage for the first row
- Copy this cell using Ctrl + C and select all the remaining cells of the column E
- Paste the contents or press Ctrl + V
- This gives all the percentages.

All the results obtained by the above steps appear in the worksheet as shown in Figure 5.7. While typing a formula for a cell, we have to start with the symbol '+'. In case of built-in functions like AVERAGE we have to start with @ symbol. Here is an interesting aspect regarding formula copy in the cells of a worksheet.

The result obtained in a cell using a formula can be copied to other cells in an easy way without using Ctrl + C and Ctrl + V method. What all we have to do is to put the cursor carefully at the bottom-right corner of the cell to be copied. We then find a '+' mark. Hold it and drag it to the other cell to which we wish to copy the formula result and leave the cursor. This will finish copying. We should note that copying a formula gives the corresponding results in the other cells and not simply copying the cell content as it is, to the other cells.

Finding the row-sums and column-sums of a table is a normal requirement. Suppose the sum of all values in a row is required. What all we have to do is to click on 'Σ' icon in the standard menu that appears on the top of the screen. The sum of all values to the left of the cell where the cursor is positioned will be given with this click. If we place the cursor at the bottom of the data in a column, the sum of values above the cell in which the cursor is placed will be given. Thus the row and column sums can be found very easily. After getting a row sum (or column sum), the same may be copied to the other cells if required. These are called the *worksheet functions* and are discussed in Section 5.8.

5.6 BUILT-IN FUNCTIONS FOR QUICK USE

There are many built-in functions in Excel to perform special calculations without typing the formula. We have to select the required function and specify the range of values on which it should be applied. These functions are known as Paste functions. In order to see the function we have to click on the icon f_x or select the buttons

Insert ▸ Function

This gives a menu screen for Paste function operations as shown in Figure 5.8.

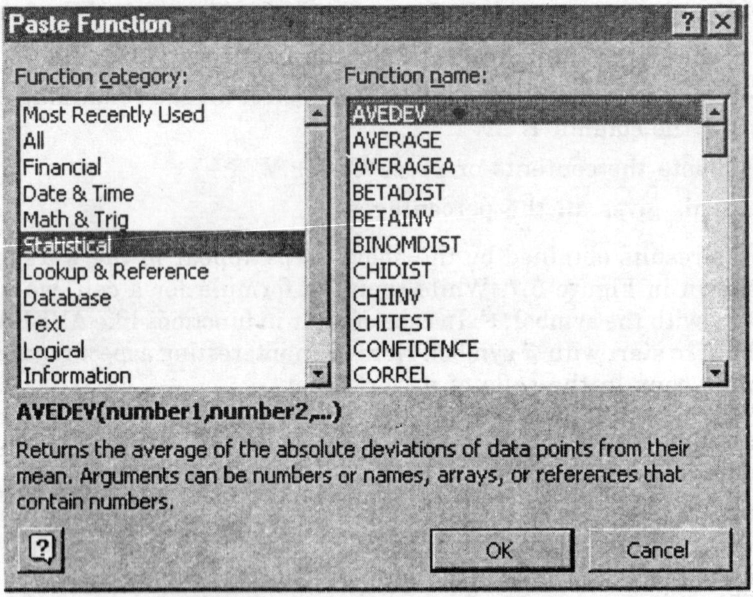

Fig. 5.8 The Paste function menu.

There are two sub-windows in this screen with a facility to select the items with mouse. The categories of functions that are available are given in the left window. After selecting the category the list of available functions is displayed in the right window. For instance, if we choose the category Financial in the left window, we get a list of financial functions in the right window. If we click on PV (present value) we get a screen asking for various input parameters like the rate of interest, payment duration, etc. The PV will be shown in the cell where the cursor is placed. The result also appears on the dialogue box.

Apart from the financial functions we have date and time functions, mathematical functions, database functions and so on. A user may need them according to the context. Since the focus is on statistical analysis, we look into the details of statistical functions only. The following is the list of statistical functions available. More details about these functions can be seen by clicking on

Help ▸ Contents ▸ Functions

AVERAGE	GEOMEAN	NORMDIST	STANDARDIZE
BINOMDIST	GROWTH	NORMINV	STDEV
CHIDIST	LARGE	PEARSON	TREND
CHITEST	LINEST	PERCENTILE	TRIMMEAN
CONFIDENCE	LOGEST	PECENTRANK	TTEST
CORREL	LOGINV	POISSON	ZTEST
COUNT	LOGNORMDIST	PROB	VAR
COUNTA	MAX	QUARTILE	
FORECAST	MEDIAN	RANK	
FREQUENCY	MINIMUM	SKEW	
FTEST	MODE	SLOPE	

Suppose we wish to find the standard deviation of production in the file FGRAINS. We can do it by selecting any cell in which we wish to display the result, clicking on the icon f_x and then selecting STDEV from statistical functions. A dialogue box appears as in Figure 5.9 for checking whether the range of cells is chosen correctly or not.

Since the first row contains the headings (field names) of the variables, the calculations have been made only from C2 to C21. If we click on OK button, the standard deviation of the values in the column C will be displayed at the cursor position. Suppose we wish to find the standard deviation of the remaining two columns. It is not necessary to repeat this procedure. It is enough to copy the present cell (by using Ctrl + C) and paste it to the two cell on the right side of it.

Instead of typing the range of cells in the horizontal window shown in Figure 5.9, we can specify the range more conveniently as follows.

- Click on the window where the input cells are to be typed.
- The cursor shifts to the data sheet from the present window.
- Select the required cells and block them with mouse. This blocking does not appear in black colour as done normally, but dotted lines appear.
- Press OK.

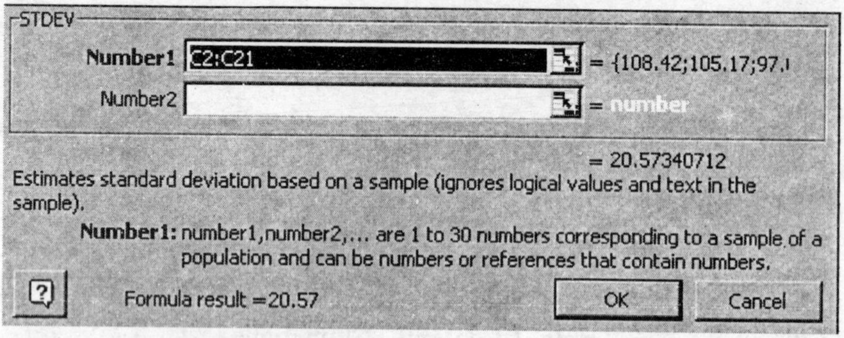

Fig. 5.9 The Paste function dialogue box for calculating the standard deviation.

This method is very useful because we need not type the address of starting and ending cells. The entire group of cells for which we need calculations can be blocked with mouse.

5.7 WORKING WITH THE TRIBAL FOOD PROBLEM IN EXCEL

Let us read the file TRIBAL.DBF that was created in Chapter 3. This file is located in C:\STATMAN. In order to open this in Excel we first select File option and click Open. Then select the directory named STATMAN from the list of folders displayed. Now select the file type as 'Dbase Files' in the dialogue box. Then the list of DBF files found in STATMAN will appear, from which we can select the file TRIBAL. This will open the file in Excel as workbook. Initially the columns may be of smaller width depending on the number of digits used in each column. The headings will also be not clear. To see the actual headings we have to select

Format ▸ Columns ▸ Autofit Selection

Now the headings will appear in full and data file looks as a big worksheet as shown in Figure 5.10.

Let us do some basic statistical calculations using the built-in functions. Suppose we wish to compare the average BMI for the three groups namely Sugali, Yanadi and Yerukala. Since the data is mixed for all the three groups under the column Code we have to filter the data before calculating the averages. By using the command Filter we will be deleting the undesired cases of the data and keep only those records satisfying the given conditions. To get Filter, the command sequence is

Data ▸ Filter ▸ Auto Filter

A mark appears on each of the columns indicating that it can be filtered if desired. Since we wish to select the tribal groups according

DATA HANDLING IN EXCEL

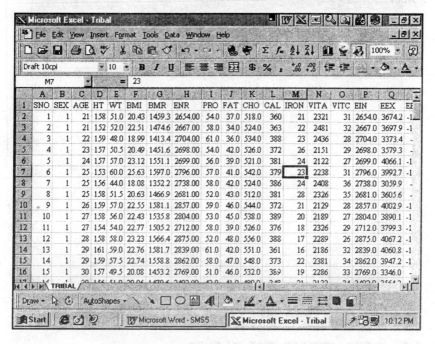

Fig. 5.10 FoxPro data opened in Excel.

to the code, we filter on this column. When we click on the column Code we get a small pull-down menu asking for the criterion for filtering.

We can click on '1' so that all records relating to CODE = 1 are selected and others disappear. We can also use Filter to select more than one criterion like CODE = 1, SEX = 1 and AGE > 30. To do this, we first filter on CODE = 1 and then filter on SEX = 1. Now we select the column AGE and chose the 'custom' method for filtering, under which we select the variable to be greater than or equal to 30. After clicking on OK we get the sorted list which contains only 12 records meeting all these conditions as shown in Figure 5.11.

Since the data has been filtered it can be copied to another worksheet for further use. To do this, copy the filtered data by selecting the required area and use Ctrl + C. If a new sheet is not available, first select Insert and then select New Sheet. This gives a new worksheet named as Sheet1. On this worksheet the cursor will be positioned in the first row in column A. Now press Ctrl + V. This would copy the filtered data to the new sheet. This file can be saved as a new workbook in Excel and not as a dBASE file. If we attempt to save the file as it is, Excel prompts a message that only the active sheet can be saved.

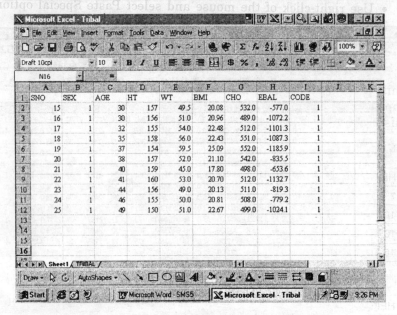

Fig. 5.11 Filtered data for the TRIBAL.

5.8 OPERATIONS ON TABLES

We can select a block of cells in a worksheet, cut them and paste them in a different worksheet or in a Word document. We proceed as follows:

- Select the cells to be blocked with mouse.
- Choose Edit ‣ Copy or press Ctrl + C.
- Select the cell within the sheet or in a new sheet where the contents should be copied.
- Choose Edit ‣ Paste or press Ctrl + V.

This is a simple operation, which is very useful for transferring the Excel output to other work areas. Here is something special called the Paste Special operation. This method can be used to Transpose data tables. A table in which rows and columns are interchanged is called a Transposed table. This is very useful in matrix operations and the result is called 'matrix transpose'. To do this, we have to perform the following steps:

- Select the cells to be transposed. For example the 10 cells in a row like A1..J1
- Press Ctrl + C
- Select another cell of the same sheet like A3 and put the cursor in this cell

- Use right-click of the mouse and select Paste Special option
- One of the options in this item is Transpose. Click on it
- The 10 elements of the previous row (horizontal) now appear as a column (vertical).

This procedure can be adopted for transposing large tables for improving the appearance of a research report. Figure 5.12 shows this effect with the help of a sample table.

SAMPLE DATA BEFORE TRANSPOSE

HEIGHT	STANDARD	RURAL	URBAN
7 TO 9 YRS	126.4	130.2	123
10 TO 12 YRS	142.7	141.96	140.13
13 TO 15 YRS	158.3	147.8	148.1
16 TO 18 YRS	163	150.81	148.33

SAMPLE DATA AFTER TRANSPOSE

HEIGHT	7 TO 9 YRS	10 TO 12 YRS	13 TO 15 YRS	16 TO 18 YRS
STANDARD	126.4	142.7	158.3	163
RURAL	130.2	141.96	147.8	150.81
URBAN	123	140.13	148.1	148.33

Fig. 5.12 The Transpose operation in Excel.

When we copy the table from Excel to Word the contents may have to be slightly edited. For instance, in the merged cells the Excel may not be in the same format in Word as they were in Excel. These things can be done only for presentation and while doing calculations in the Excel sheet; no editing is necessary. Suppose we wish to find the inverse of a 3×3 matrix given as

$$A = \begin{bmatrix} 2 & 3 & 6 \\ 1 & 4 & 8 \\ 0 & -1 & 4 \end{bmatrix}$$

The following steps are sufficient to do this job.

- Enter the data in the worksheet, in 3 rows and 3 columns say from B5 to D7.
- Move to another location in the same sheet and block another 3 rows and 3 columns, say D5 to H7.
- Select MINVERSE from the paste functions and block the cells B5 to D7 as input in the dialogue box.
- Press Ctrl + Shift + Enter.

- The inverse is displayed in the cells D5 to H7 denoted by the matrix B.
- The matrix A and the inverse matrix B are shown in Figure 5.13

	A	B	C	D	E	F	G	H	I
1									
2			GIVEN MATRIX IS - A				INVERSE - B		
3									
4		2	3	6			0.8	-0.6	0
5		1	4	8			-0.13333	0.266667	-0.33333
6		0	-1	4			-0.03333	0.066667	0.166667
7									
8									
9			MATRIX PRODUCT - A*B				1) If the matrix is singular, an error		
10							message will be displayed		
11		1	0	0					
12		-5.55112E-17	1	0			2) Product of a matrix with its		
13		2.77556E-17	-5.55112E-17	1			inverse is the identity matrix		
14									

Fig. 5.13 Matrix inverse and multiplication.

We can also multiply two matrices by choosing the paste function MMULT and selecting the two arrays (matrices) to be multiplied. Let us multiply the matrix A with its inverse B. The product should be the identity matrix in which the diagonal elements are all '1' and the other elements are '0'. Following the steps in the dialogue box, we get the product as shown in Figure 5.13. The elements in the product matrix on the lower side of the diagonal are very small numbers which can be taken as '0'.

5.9 PRINTING THE DATA AND RESULTS

Sometimes we need to print the data from the worksheet. To do this, select the data area to be printed by blocking the area and press Ctrl + P. A screen with several options for printing appears. Among them we find an option 'Selection' and Excel automatically chooses it for printing. If we want the entire sheet to be printed we select the option 'Entire Worksheet'. When we click on OK or press the Enter key the selected sheet goes to printing. We have to, however, ensure that the printer is connected properly.

There are several ways of getting the output. Before we print, it is advised to view the sheet by selecting the option Print Preview which is found in the File options. This shows the sheet on the screen just as it would appear in print. The Print Preview has several options as under:

- If we select the Setup option, we find a choice as to whether the sheet be printed as Portrait (vertically on the paper) or Landscape (horizontally on the paper). We can do this depending on the size of the sheet and the orientation required.
- We can also avoid printing the Grid Lines (the lines, which appears on the sheet) by not selecting that option. This would save a lot of printing effort. This choice is found in the option named as 'Sheet'.
- We can see the margins seen in the preview and change them, if necessary, by dragging them to the desired position with the help of the mouse.

Whenever required we can block a few columns of the worksheet and print those columns only. When there are too many columns the output will be split into more than one page. To see this effect, we click on Print Preview. If the output goes to more than one page, we find a highlighted menu item called Next Page on the top left corner of the screen. Depending on the paper size, the number of columns and the rows that can fit into a page will be determined by Excel automatically.

We can adjust the column width by holding the right border of the column and dragging it to the right or the left. We can reduce or increase the number of decimals in the cells by choosing the corresponding item in the main menu. This works only for numerical data. With all these alterations we can adjust the output of Excel to some extent so that it may fit into a page.

We close this chapter with these basic details on data handling in Excel and proceed to the construction of graphs and charts using Excel, in the next chapter.

REFERENCE

1. Microsoft Office—Excel reference from online help.

DO IT YOURSELF

5.1 Open an existing Excel worksheet and right click on the label Sheet1, which is found at the bottom of the work area. A small menu appears with options for renaming or deleting a file. Try to delete the current sheet and see what message Excel gives? If you do not want to delete, click on NO.

5.2 Click File ▸ Open and examine the different types of files that can be opened in Excel.

5.3 Click View and select Toolbars. Study carefully the different toolbars that can be activated. Do you think that all of them are necessary always?

5.4 Open a new worksheet and start entering the following fields in the first row.

DATE   SALES   UNIT   PRICE   REMARK

By selecting the options Format ▸ Cells we can define the nature of the data to be entered in the cell. If we choose the SALES field as # # # we can enter data with 3 digits. Like this, try to create your own file with different cell types.

5.5 Consider any existing Excel worksheet. Insert a column and drag the data of another column to this column. This is necessary to bring the selected columns of a big worksheet to adjacent positions before creating graphs or performing certain statistical routines.

5.6 Create an Excel worksheet and save it as

(a) dBASE file with file name C:\MYDOCUMENTS\TRY1

(b) Text file with file name C:\MYDOCUMENTS\TRY2

5.7 Open an Excel worksheet. Copy a portion of its contents to Sheet2. Now try to save it as a DBF file. What message do you notice?

5.8 A company has employed handicapped persons along with others in its workshop. After an elapsed time of 45 days, the performance of these employees has been assessed and the following ratings were given:

Table 5.1 Ratings of Employees

Employees	Performance		
	Above average	Average	Below average
Blind	21	64	17
Deaf	16	49	14
No handicap	29	93	28

(a) Create a worksheet in Excel for this data along with headings. (*Hint:* Columns A, B, C, D, E to be used with row 1 and row 2 for headings.)

(b) Find the row totals, column totals and grand totals by creating new headings for this purpose. Calculate the %

of blind, deaf and no-handicapped employees in the total employees. (*Hint:* Use +(F3 * 100/F6) for column % and enter. Then copy to all other rows.)

5.9 Block cells of the data in Problem 5.8 and print it. (*Hint:* Block cells ▸ Ctrl + P. Follow the dialog box.)

5.10 Open C:\STATMAN\TRIBAL.DBF in Excel. Block all columns and click

Column ▸ Autofit Selection

Now use Print Preview. Check, how many pages the output needs.

Graphs and Charts in Excel

Excel has a built-in facility for creating graphs and charts. There are several types of graphs supported by Excel. Though all of them are useful for presenting statistical data, we shall discuss *bar charts, line charts, pie charts* and *scatter diagrams*.

To invoke the graphs in Excel, it is enough to click on the graph icon from the menu bar. A big menu with several types of model graphs appears on the screen as shown in Figure 6.1. By default, the bar chart is always ready for use.

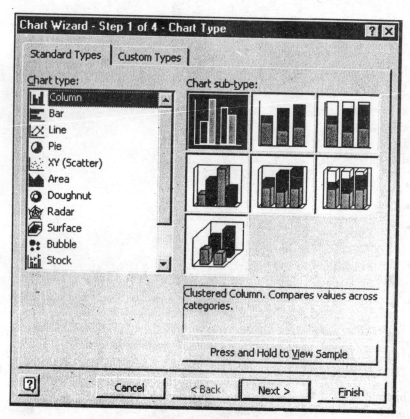

Fig. 6.1 Dialogue box for selecting the chart.

6.1 CONSTRUCTION OF A BAR CHART

A bar chart is constructed to show the frequency of a categorical variable. The horizontal axis contains the categories and the vertical axis is used to show the number of cases (frequency) of the characteristic in which we are interested. For instance, we can use the bar chart to show the yearly production of a company with respect to one or more products. Consider Example 6.1.

EXAMPLE 6.1 The production of wheat (in tonnes) in different years in a particular region are given as in Table 6.1:

Table 6.1 Production of Wheat (in Tonnes)

Year	Production of wheat
1980–81	6124
1981–82	7004
1982–83	7216
1983–84	7060
1984–85	6950
1985–86	6916
1986–87	7620

To construct a bar chart for these data, we first enter the data in a worksheet of Excel as two columns with headings in the first row. We can as well enter the same data in two rows with headings in the first column. After the data is entered, the following steps may be adopted to construct the chart:

- Click on the graph icon.
- Click Next.
- Fill in the details of the heading for the chart and the titles for the X-axis and the Y-axis.
- Within the dialogue box, several options are available for choosing the gridlines, legend, title, etc.
- After finishing this stage click on Next. Excel asks whether the chart is to be saved in the same worksheet where the data is available or as a separate worksheet.
- Click on Same Sheet so that the data and graph appear on the same sheet.

The bar chart for this data is shown in Figure 6.2. This graph can be pasted in any document using the cut-and-paste operation. ∎

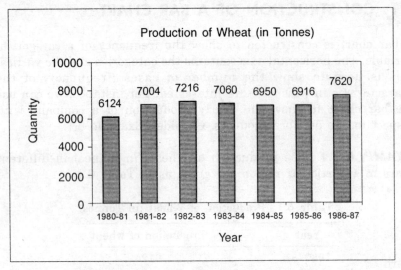

Fig. 6.2 Bar chart for the production of wheat.

6.2 CHART OPTIONS

After choosing the type of chart we can select several properties of the chart and make it understandable and elegant. The chart contains two broad areas: *plot area* and *chart area*. The plot area is usually shown by a shade or colour, whereas the chart area is the entire space embedded in a frame in which the plots of the graph, the legend, the title, etc., are present. If we right-click on the chart area we find several options like Chart Type, Location, etc.

We can click on the Chart Type to change the type of the graph from bar chart to line chart or to scatter diagram or to a different type of bar chart. Steps are as under:

- Click on Chart Options to select various aspects like Title, Legend, Grid Lines and so on.
- Click on Location to change location of the graph from the Same Sheet to a New Worksheet. Excel will show it as Chart-1 and we may change the name, if necessary. It is a good practice to name the chart in such a way that it can be easily identified.
- We can avoid Grid Lines by a right click on any Grid Line. A dialogue box appears in which one of the options is Clear. If we choose this option, we can avoid grid lines. This could also be done while constructing the chart, in the process of giving title, legend, etc., where the dialogue box asks for the Grid Lines to be shown or not.

- We can now think of Legend, which is an indicator of various lines/bars and their styles on the chart. Usually this is shown on the Right of the plot area but we can change its position by a right-click on the Legend and act as per the choices given in the dialogue box.
- We can choose the data value to be shown on the bar itself. Right clicking on Any One Bar and choosing one of the options given therein can do this. One of the options is Show Value and this shows the value of the Bar on the top of the bar.
- We can choose Font to select the type, the size and the alignment of various aspects like Title, Data Labels, Legend and so on. The choice of the Font is important since the size of the letters should match with the size of the chart.

With these choices the bar chart appears as shown in Figure 6.2. We note that in this example, the data given under YEAR contains a non-numeric character '-' as in 1980-81. Excel always takes the *first* column of the worksheet on the *category axis*. This axis is used to show various categories of the variable under consideration like male and female or tribal groups like Sugali, Yanadi, Yerukala and so on.

Consider another example of a company, which makes tables and chairs.

EXAMPLE 6.2 The production of tables and chairs by a furniture company are shown as Table 6.2:

Table 6.2 Production of Tables and Chairs

Year	Tables	Chairs
1990	185	146
1991	242	182
1992	169	254
1993	126	188
1994	96	112
1995	148	162
1996	176	242
1997	194	286

Let us construct a bar chart to show the production of tables over different years. If we enter the data as a worksheet and block the first two columns and proceed as in the case of Example 6.1 we get the bar chart shown in Figure 6.3. If we select the column *chairs* also we get a multiple bar chart. ■

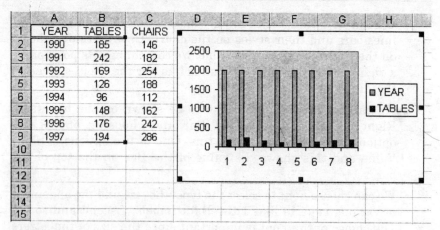

Fig. 6.3 The bar chart due to incorrect choice of the chart type.

What is wrong with this chart? On the category axis, instead of YEAR we have got the serial number. The YEAR is shown as one set of bars and the TABLES is shown as the second set of bars. This is certainly not what we wanted to do. The reason for this should be watched closely. When the first column contains numeric data as in this example, Excel starts with that column for plotting and searches for data to be shown on the category axis. As the first column itself is a numeric data, Excel takes the serial number on the category axis.

In order to overcome this problem, we have to give data as non-numeric under the heading YEAR. Since the data is already typed, we need not be concerned a lot. Choose Chart options and select Scatter Diagram. The diagram appears with dots instead of bars. We now see that the category axis contains the year and not the serial number. Again choose Chart options and select the Bar Chart option. Then the chart appears as shown in Figure 6.4.

We can change the properties of the bar with a right click on any bar. We find a menu with title Format Data Series. By clicking on this title we get a dialogue box with the following items and facilities.

Patterns

By clicking on this we get a variety of colours/shades and different patterns like dots, stripes, etc., to fill in the bars.

Axes

By selecting the item Scale we can modify the minimum and maximum values of the Y-axis. The major and minor divisions can also be selected as desired.

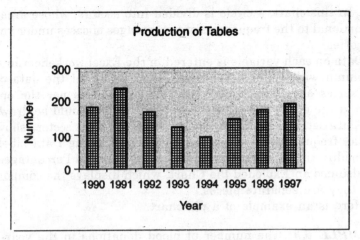

Fig. 6.4 The bar chart after choosting scatter diagram and then clicking the bar chart option.

Y-bar errors

Error bars are used to express the potential error in plotting the values relative to each data marker in a data series. Under this option we can select either one or two standard deviations or the standard error of the data series as Y-BAR ERROR. We can alternatively use a custom designed error by choosing the plus '+' or minus '−' components around the value.

Data labels

We can choose data labels to be shown against each bar. We can select the value or the label or both.

Series order

In case of a multiple bar chart, we can alter the order in which the bars should appear corresponding to the data series.

Options

This is an important application where we can select the gap width of the bar from its size 150 (by default) to any other size. If we make it zero the bars will be attached to each and the bar chart looks like a histogram. We can also allow overlap of bars by choosing the amount of overlap. By default the overlap is set as zero.

6.3 CONSTRUCTION OF A PIE CHART

A *pie chart* is used to represent the distribution of a categorical data. In this chart, a circle is divided into sectors, whose areas are proportional to the frequencies or percentages of cases under various categories.

Data on each variable is entered in the Excel worksheet in a row or column with suitable headings. Then we select the data of the first series and choose pie chart and proceed as per the options given in the dialogue box. A separate pie chart should be drawn for each data set. The pie chart appears with different sectors shown as per the frequencies given in the data. By choosing Data labels we can modify the label on the sectors from Value to Percentages. We can also use an Exploded Pie Chart, which displays the components as if they are separate pieces.

Here is an example of a pie chart.

EXAMPLE 6.3 The number of blood donations in the year 1995 according to various groups is as follows (data source: File—blood.dbf):

Table 6.3   Number of Blood Donations

Year	Blood group			
	O	A	B	AB
1995	1154	526	775	155
1996	1250	620	780	200

Let us show the data for the year 1995 in a pie chart. To do this, the data is first entered in Excel worksheet and blocked with the mouse.

By choosing the pie diagram option with data labels to be shown as percentage and the legend at the bottom, we get the pie chart as shown in Figure 6.5. Each time, only one row of data can be used to draw a pie chart. ■

We can now choose other types of pie chart or change the chart options on the existing chart. The same chart can be shown as *exploded* by changing the chart type. This gives the diagram shown in Figure 6.6.

For each set of data, a separate pie chart is to be constructed and can be assembled on single sheet. We can alternatively use *doughnut* chart from the chart menu to show multiple pie charts. A *sub-divided bar diagram* as shown in Figure 6.7 can alternatively show the data given in Example 6.3.

6.4 CONSTRUCTION OF A LINE CHART

Line charts are used to represent time–series data. The category axis should be non-numeric like month or year. If it happens to be a numeric one, the line chart plots these values also as another response variable. We can simultaneously plot more than one series on the same graph. Here is an example.

GRAPHS AND CHARTS IN EXCEL

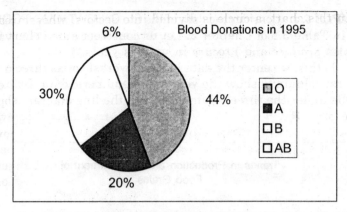

Fig. 6.5 A pie chart.

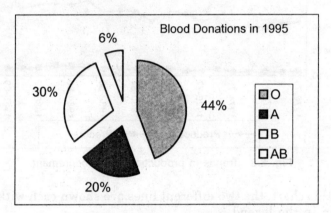

Fig. 6.6 An exploded pie chart.

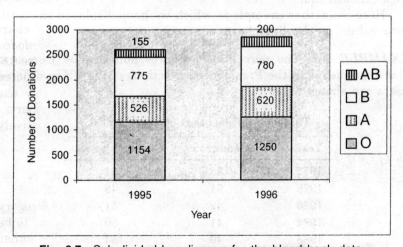

Fig. 6.7 Sub-divided bar diagram for the blood bank data.

EXAMPLE 6.4 Consider the data of the file C:\STATMAN\FOOD.DBF given in Table 3.2 of Chapter 3. Let us construct a line chart for the variables *produce* and *procure* in various years.

To do this, we enter the data in Excel worksheet as three columns with first column to show the year as non-numeric data. By blocking the cells under the three columns, we get the line diagram shown in Figure 6.8. ■

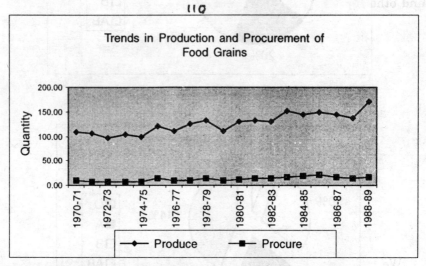

Fig. 6.8 Trends in production and procurement.

In this chart, the two different lines are shown each with a style as given in the legend.

The chart title and the size of the font can be selected by a right click on each line.

Here is another example in which we have displayed the actual data values on the line chart.

EXAMPLE 6.5 The percentage of votes obtained by Democrats and Republicans in the Presidential Elections of the US in different years are shown below:

Table 6.4 Percentage of Votes Polled

Year	Democrates	Republicans
1972	38	61
1976	51	48
1980	42	51
1984	41	59
1988	46	54
1992	43	38

Let us depict these figures by a line diagram and fix the data values on the chart itself.

First we enter this data on the Excel worksheet. If we have typed this data in Word, we can simply block the cells of this table, copy and paste in the Excel worksheet. We need not again type the data.

Now, we select the line chart as usual and fix the titles, legend and other options. The chart now appears as shown in Figure 6.9.

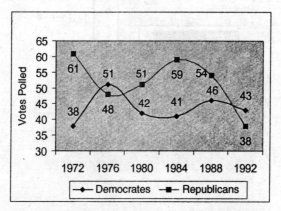

Fig. 6.9 Trends in voting over different years.

We have shown the data values on the chart itself. If we right-click on one line at any place, we get a menu with the heading Format Labels as shown in Figure 6.10.

We have chosen to show the data value near the point either above or below the line. This option is available with the item Alignment in the dialogue box as shown in Figure 6.10. We can also change the style of the label from horizontal to vertical position.

Since we have two data series, the pasting of data values has overlapped at one or two points. By identifying such points, we can pull down or pull up the data value by holding it with mouse and moving it up or down.

Apart from the standard models for charts we have a series of other charts called *custom type* charts. We can view them by clicking on this button in the chart menu as shown in Figure 6.1.

6.5 CUT, COPY, PASTE AND PRINT OPTIONS FOR GRAPHS

We can copy the graph from Excel sheet to another document in Word or to a PowerPoint presentation (slide) by using the cut and paste operation. Suppose a report requires data and the graph to be shown in the same document. To do this, we perform the following steps:

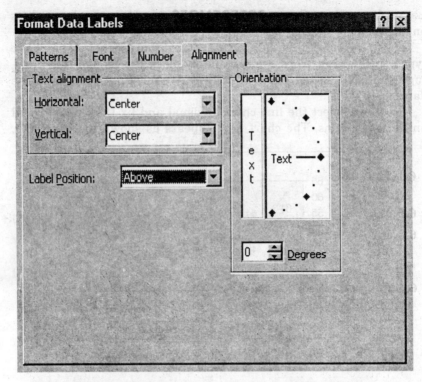

Fig. 6.10 Dialogue box for alignment of data labels.

- Right-click on the graph
- Choose Copy
- Now select Word document and locate the place where to paste the graph
- Again right-click in the document
- The Paste option alone will be highlighted
- Click on Paste
- The graph gets pasted

In order to print a graph, we first choose the Print Preview option. The graph appears in one of the two forms Portrait or Landscape.

If we choose Landscape the graph is printed horizontally on the paper. Another way of printing the graph is as a Portrait, which shows the graph vertically. Print Preview is therefore useful for viewing the graph before it is printed.

Thus Excel graphs are very useful for common user requirements. There are other software packages that help in creating special type of graphs.

REFERENCES

1. Microsoft Office 97 reference from online help.
2. Robert Cowart (1995): *Mastering Windows 95—The Windows 95 Bible*, BPB publication, New Delhi.

DO IT YOURSELF

6.1 List out the different types of charts available under Custom Type and see what are various applications of these charts.

6.2 Find out how a three-dimensional bar chart can be created.

6.3 Is it possible to give different shades to the bars after the chart has been prepared. Try with a right click on any bar and check the options.

6.4 Construct a subdivided bar diagram for the following data showing the amount spent on various heads of expenditure.

Year	Food	Education	Others	Total
1988	300	200	250	750
1989	350	300	375	1025
1990	400	350	425	1175
1991	500	500	550	1550

6.5 The following is a frequency distribution of data obtained by a user. Construct a frequency curve. (*Hint:* Enter the values of the class intervals as 10–20 by giving a space before and after '-'. Then select the line chart.

Class	Frequency
10–20	4
20–30	8
30–40	12
40–50	15
50–60	11
60–70	7
70 & more	3

6.6 The percentage rate of return (RR) on money market investments during various months of a particular year, worked out by a financial expert is given below.

Month	RR %	Month	RR %
Jan	6.2	Jul	6.0
Feb	5.8	Aug	6.8
Mar	6.5	Sep	6.5
Apr	6.4	Oct	6.1
May	5.9	Nov	6.0
Jun	5.9	Dec	6.0

Construct a suitable chart.

6.7 What options do you select if the data table is also required along with the chart?

6.8 Suppose we have constructed a bar (or a pie) chart using rows of data. We now wish Excel to change the data reading from rows to columns and modify the chart. How can this be done?

6.9 Click on any bar and examine the facilities for changing the shades and fill pattern of the bars. This is available in Format Data Series option.

6.10 What is the use of the option Format Plot Area? Check whether you can change the background to the plot area from one shade to another.

Descriptive Statistics Using Excel   7

> *Drawing conclusions and understanding more about the sources of our data is the goal of statistical analysis.*
> —*Kachigan*

The observations of the real world are converted into numbers by the researcher. The numbers are manipulated and organized so as to yield a summary of the sample findings. The computer has come handy in doing this part of the job effectively and in short time. The next job is to interpret the results and translate them back to the real world so that they are better understood than how they were prior to data analysis.

In this chapter, we discuss some analytical aspects that can be carried out with the help of MS-Excel.

7.1 DATA ANALYSIS PAK IN EXCEL

Excel has a built-in statistical package for carrying out data analysis. This item is usually hidden and can be brought out to the menu by clicking the buttons Tools ▸ Add-Ins.

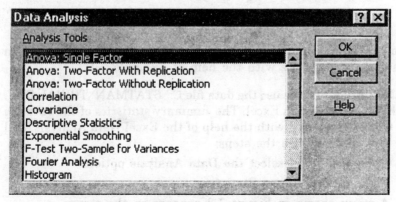

Fig. 7.1 Statistical analysis tools in Excel.

We then get a menu as shown in Figure 7.1. Many of the commonly required statistical tools are included in this section of Excel. It is therefore worth examining the meaning and interpretation of the different items in this menu.

7.2 DESCRIPTIVE STATISTICS

The first broad function of statistics is to describe the data in an interpretable form. This can be done either in terms of numerical values called *summary statistics* or by showing the data as graphs. The ultimate objective in this part of statistics is to describe the data in a proper way. The Data Analysis Pak has specific tools to perform this job.

7.2.1 Summary Statistics

Under this section all the estimates of the measures of central tendency, dispersion, shape and the standard error of the mean are given as output for the selected variable (shown under one column). It is important to note that we are always estimating the unknown population parameters using the sample data. So, while reporting the results it is necessary to specify the standard error of the estimates.

The different summary statistics available in the Data Analysis Pak are shown as in Figure 7.2.

Mean	Range
Standard Error (of the mean)	Minimum, Maximum
Median	Sum
Mode	Count
Standard Deviation	Largest value
Variance, Kurtosis	Smallest value
Skewness	Confidence Level

Fig. 7.2 Summary statistics available in Excel.

Let us examine in detail the method of using this facility to obtain the summary statistics with the help of Example 7.1.

EXAMPLE 7.1 Let us use the data file C:\STATMAN\TRIBAL.DBF, which can be opened in Excel. The summary statistics of the variable BMI will be computed with the help of the Excel Data Analysis Tool Pak. The following are the steps:
- Open Excel and select the Data Analysis option.
- Click on Descriptive Statistics.
- A menu shown in Figure 7.3 appears on the screen.

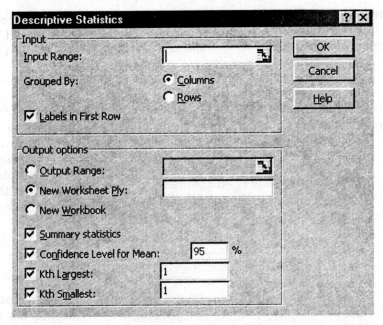

Fig. 7.3 Menu window for input–output selection.

- There is a small window against the item Input Range. Click on with mouse in this window.
- It is a good practice to click on the item Labels in First Row so that the output shows the name of the variable for which the results are obtained.
- The cursor shifts from this screen to the data screen. With the help of the mouse, select the column F in which we have the data on BMI. We find the selection is not in black colour as happens in general, but it appears as a dotted line along the columns.
- Choose the output options by selecting them with the mouse. By default, the output appears as a separate worksheet, called *new worksheet ply*.
- Now click on OK. The output appears as shown in Table 7.1.

Table 7.1 Output of Descriptive Statistics

Mean	19.930	Range	11.770
Standard Error	0.173	Minimum	14.600
Median	19.750	Maximum	26.370
Mode	20.700	Sum	2989.480
Standard Deviation	2.121	Count	150.000
Sample Variance	4.499	Largest(1)	26.370
Kurtosis	0.810	Smallest(1)	14.600
Skewness	0.579	Confidence level(95.0%)	0.342

The output actually appears as two columns, which we have divided here into four columns for convenience in presentation. The contents of the above table are self-explanatory. It contains all measures of central tendency and those of dispersion. The new items that appear in the list are standard error and confidence level (95%). The *standard error* (SE) given above is a measure of the amount of error in the estimation of the population mean based on the sample data. Smaller the SE, more reliable is the estimate. So, it is a good practice to present the SE whenever the mean is calculated. Consider Example 7.2.

EXAMPLE 7.2 Two independent investigators A and B have collected the following sample values on the hardness values of rubber balls (Table 7.2). We wish to estimate the average hardness value of a ball and compare the values obtained by the two investigators.

Table 7.2 Results by Two Investigators

A	B	A	B
180	178	300	256
140	120	270	240
230	150	110	92
190	175	240	200
160	129	320	150

Using the Data Analysis Pak, we get the summary statistics as shown in Table 7.3. If there are no duplicate values in a data series, Mode will be denoted # NA, to mean *not available*.

We find from these results that the mean is estimated as 214 by the investigator A, while the estimate from the data of B is only 169. When we look at the standard error of these two estimates, we find that B has a lower standard error than that of A and hence it is reasonable to consider the estimate provided by B rather than that by A.

Table 7.3 Summary Statistics from Excel

A		B	
Mean	214.000	Mean	169.00
Standard Error	22.020	**Standard Error**	16.45
Median	210.000	Median	162.50
Mode	# NA	Mode	150.00
Standard Deviation	69.634	Standard Deviation	52.00
Sample Variance	4848.889	Sample Variance	2704.44
Kurtosis	−1.105	Kurtosis	−0.54
Skewness	0.117	Skewness	0.39
Range	210.000	Range	164.00
Minimum	110.000	Minimum	92.00
Maximum	320.000	Maximum	256.00
Sum	2140.000	Sum	1690.00
Count	10.000	Count	10.00

Referring to Table 7.1 we find the 95% confidence level as 0.342. The confidence level is used to construct an interval around the estimated mean in such a way there is a 95% chance that the true mean lies within this interval. This is called *confidence interval*. The formula to determine the confidence interval for mean is

$$\bar{x} \pm 1.96 \frac{\sigma}{\sqrt{n}}$$

where $\bar{x}$ is the observed mean and σ the standard deviation. The confidence level given by Data Analysis Park is the amount to be subtracted from and added to the mean $\bar{x}$. This is based on the assumption that in a normally distributed data, about 95% of the data values lie within 2σ limits from the mean. In the case of BMI data, the confidence interval will be 19.9299 ± 0.3422 which is equal to (19.5877, 20.2721).

7.2.2 Frequency Distribution and Histogram

The data analysis park has a facility to prepare the frequency distribution and the histogram for the data on a single variable at a time. This is based on the raw data entered in the worksheet and Excel creates a frequency table by taking the class width automatically. However, if we wish to specify the class intervals in a particular form, we can type the CLASS UPPER LIMITS called Bins in a column and supply that information as input while working out the histogram. The output has different options as shown in Figure 7.4.

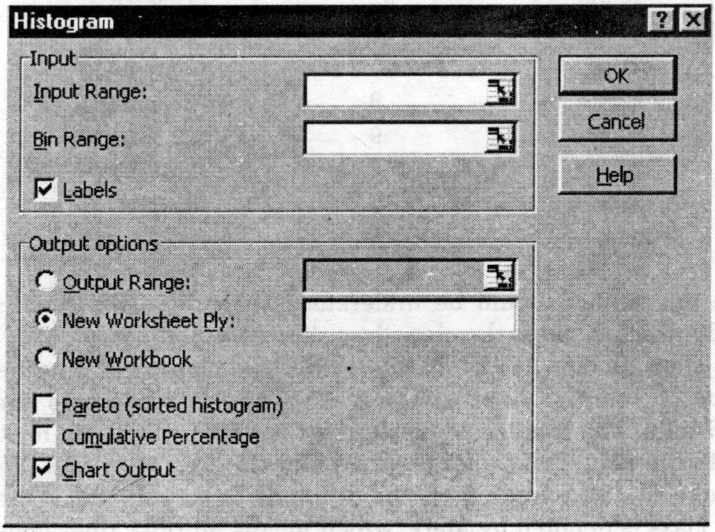

Fig. 7.4 Histogram options.

We can also choose a sorted histogram, which shows the frequencies in the decreasing order. This is called the *Pareto diagram*. It is used to identify the *vital few* and the *trivial many* items of the data.

Consider Example 7.3 in which the histogram is constructed with automatic bin intervals.

EXAMPLE 7.3 Reconsider the TRIBAL file. Let us construct the histogram for the variable BMI. Using the tool Park select the histogram option. As usual the input range can be specified by clicking on the empty window and then blocking the values of BMI in the worksheet along with the heading. We should not forget to tick the item *Labels in the first row* failing which, an error message appears that a non-numeric value is included in the data.

When it comes to Bin ranges let us first leave the range unspecified. Then Excel automatically selects a *group of evenly spaced bins* over the data range. From the output options, select the Chart Output. The output shows both the frequency distribution and the chart on the same sheet. Let us look at the frequency distribution of BMI shown in Table 7.4.

Table 7.4 Frequency Table of BMI Data

Bin	Frequency	Class
14.60	1	Less than 14.6
15.58	1	14.6–15.6
16.56	4	15.6–16.6
17.54	6	16.6–17.6
18.52	25	17.6–18.6
19.50	32	18.6–19.5
20.49	28	19.5–20.5
21.47	25	20.5–21.5
22.45	9	21.5–22.5
23.43	9	22.5–23.4
24.41	4	23.4–24.4
25.39	2	24.4–25.4
More	4	25.4 and more

The bin values should be understood carefully. The first value is 14.60 which denotes the class 'less than 14.60'. It contains all values less than 14.60 and the frequency of this class is 1. The next bin value is 15.58 or approximately 15.6 and it represents the class 14.6–15.6. The frequency of this class is also 1. Similarly for the bin 22.45 the actual class is 21.5–22.5 and the frequency is 9. We have to take care of rounding-off the decimals to meaningful values. The last class is denoted by *More*, which means all the values above 25.4 and the frequency is 4 for this class. It can be seen that the class intervals are not uniform. This is the case when Excel creates a 'fair' distribution of data into different groups without loosing continuity in the classes.

We can rewrite the bins into classes and understand the table in the usual way. These classes are shown in the third column of Table 7.4. The chart output first appears as a **bar** chart and that too in a small size. We have to use graphic options to convert it into a histogram. Following are the steps required to improve the appearance of the graph:

- Select the chart area and pull it down from the bottom row
- Click on legend and clear it because it is not necessary
- Click on axis and select Format Axis
- Choose Fonts and change the font to a smaller size
- Select Alignment to modify the axis values to vertical position, if necessary
- Click on the bars and select Format Data Sources. Choose Value to be shown
- Click on Options and choose the Gap Width
- Choose the width to be Zero from its original 150 (default)
- Right click on the horizontal axis
- On the resulting screen, choose the item Major Tick Marks and choose None. This avoids unnecessary tick marks on the axis
- This will automatically bring the bars close to each other and creates the histogram.

With all this exercise, we get the histogram as shown in Figure 7.5.

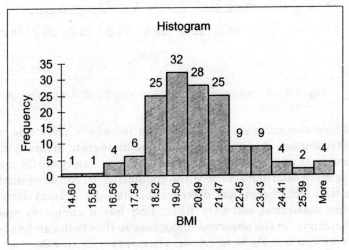

Fig. 7.5 A histogram for the data Example 7.3.

In Example 7.4, we construct another histogram with custom-designed classes.

EXAMPLE 7.4 Suppose we wish to create bins (class limits) as per our choice for the distribution of BMI values. The first thing is to find that the range of BMI values. From Table 7.1, we see that the minimum is 14.6. If we wish to take the class interval as 2 we can have 7 classes as 14–16, 16–18, 18–20, 20–22, 22–24, 24–26 and 26–28. The upper limit of each of these classes is the input information to be given for the Histogram dialogue box. Let us centre the bin ranges in the cells G153 to G160 of the sheet in the file TRIBAL. In fact we can enter them in any empty row or column and it would be useful to give the heading as BIN for identity.

We have chosen the column G because the actual data is in the column F. Now if we choose the Histogram to run, Excel asks for Bin Range to be entered in the window against the item. We simply click on it and get back to the cell G153 and block all the bin values. With a click on OK button we get another graph on the same sheet as shown in Figure 7.6, which is edited as in Figure 7.5.

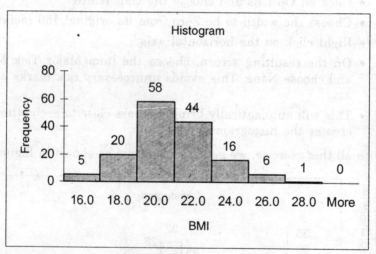

Fig. 7.6 A histogram with custom-designed class intervals.

Since the purpose of a histogram is only to show the shape of the distribution it is enough to use the histogram given by Excel for all practical purposes. A histogram becomes handy while comparing two or more distributions or while comparing an observed distribution with that of a theoretical distribution like the normal distribution. Certain statistical software like SPSS has a choice to embed the normal curve on the observed histogram so that both can be compared. Thus we can use Excel to create a histogram in an easy way. Unless we have some predefined class limits, it is suggestive to leave the bins unspecified.

7.3 CROSS-TABULATIONS AND PIVOT TABLES

We can use Excel for preparing cross-tabulations across categorical variables, which is a basic requirement in data analysis. In statistics, we call it a *two-way frequency table*. We have seen (in Chapter 3) a method of counting the frequencies belonging to two categorical variables, using FoxPro. It was a manual procedure and we have to write a program to get it done automatically. In Excel, there is a built-in facility called *pivot table wizard*. We can prepare a custom-designed table in two, three or even more dimensions with the help of this wizard. Consider Example 7.5.

EXAMPLE 7.5 Srinivas (1999) has measured the academic stress of students of intermediate classes studying in different colleges of Tirupati. The study has covered 480 students, classified as arts and science, private and management and boys and girls. One of the response variables is the total academic stress which is measured in five dimensions coded as F1, F2, F3, F4 and F5. The sum of these variables is taken as the measure of the academic stress and labelled as FTOT. A small portion of the entire data file is shown in Figure 7.7. (Due to space constraint the full file is not given here. It is saved as C:\STATMAN\STRESS.DBF and read in Excel.)

	A	B	C	D	E	F	G	H	I	J	K	L	M	N
1	SNO	SEX	AS	MGT	BORD	MOCC	FOCC	F1	F2	F3	F4	F5	FTOT	FCOD
2	1	1	1	1	1	2	3	18	15	11	19	13	76	2
3	2	1	1	1	2	2	1	22	20	21	18	20	101	3
4	3	1	1	1	1	2	3	16	6	23	9	15	69	2
5	4	1	1	1	2	2	3	5	8	7	7	9	36	1
6	5	1	1	1	1	2	4	15	15	12	18	13	73	2
7	6	1	1	1	5	2	1	13	28	16	8	13	78	2
8	7	1	1	1	1	2	4	5	5	6	4	9	29	1
9	8	1	1	1	2	2	5	12	12	10	8	11	53	1
10	9	1	1	1	1	2	4	8	9	4	3	3	27	1
11	10	1	1	1	1	2	5	8	6	3	5	4	26	1
12	11	1	1	1	1	2	4	22	17	22	30	22	113	3
13	12	1	1	1	3	2	3	24	17	20	14	18	93	3
14	13	1	1	1	3	2	5	14	15	13	15	15	72	2
15	14	1	1	1	1	2	4	14	17	26	7	15	79	2

Fig. 7.7 Part of the data on academic stress (FTOT). The file contains 480 records.

The variables and their codes found in the above table are as follows:
- SNO (Serial Number).
- SEX (1 for Male and 2 for Female).
- AS (Arts or Science: 1 for Arts and 2 for Science).
- MGT (Management Type: 1 for Private and 2 for Endowment).

It is required to make a cross-tabulation of FTOT according to SEX, AS and MGT. These three are the categorical variables, called *factors* and the response variable is the FTOT. This job is done with the following steps:

- Select the item Data from the main menu and then choose Pivot Table Report
- The source from which Excel has to get the data will be asked in the function screen. Click on Next in the resulting menu screen, if the data is in Excel file. The data selection screen appears as shown in Figure 7.8.

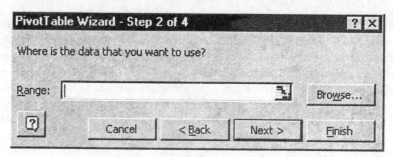

Fig. 7.8 Data selection screen for pivot tables.

Now click on the window against the item Range and then Select the columns having the required fields. Note that all the required fields shall be adjacent to each. If one field is away from the categorical variables, it has to be brought to that place by inserting an empty column and moving the required data into it.

- Now click Next. The pivot table screen appears as shown in Figure 7.9.

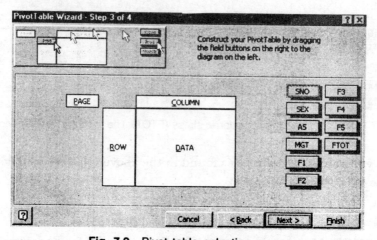

Fig. 7.9 Pivot table selection menu.

DESCRIPTIVE STATISTICS USING EXCEL

- Now drag the field to be shown along the rows.
- Drag one or more fields to be shown along the columns. In this case, we choose the variable SEX along the rows and the two variables AS and MGT along the columns.
- The variable AS will be the first classification and MGT is nested in it.
- The next job is to select and drag the response variable into the space provided for data. Here we drag the field FTOT into this location and click Next. Excel then asks whether the output be given as a separate sheet or in the same worksheet.
- Now click on New Sheet for easy handling. By default, the details in the cross tabulation will be given for Sum (of FTOT). This appears in the area earmarked as Data in Figure 7.9.
- Double click on Sum and this gives a dialogue box as shown in Figure 7.10.

Fig. 7.10 Dialogue box for Sum of FTOT.

- Select Count and then click Finish. This gives the pivot table as shown in Figure 7.11

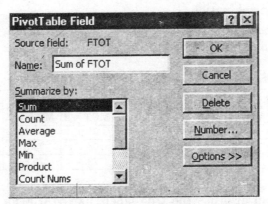

Fig. 7.11 Pivot table as given by Excel.

To understand Fig. 7.11, we have to note that the first classification along the columns is over AS and the next one is MGT. So we have got the classification for AS = 1 under which again we have MGT = 1 and MGT = 2. As a result, the number of cases falling under SEX = 1, AS = 1 and MGT = 1 is 60. This can be understood as shown in Table 7.5.

Table 7.5   Cross-tabulation of Counts of FTOT

Count of FTOT	AS = 1		Total 1	AS = 2		Total 2	Grand total
	MGT = 1	MGT = 2		MGT =1	MGT = 2		
SEX-1	60	60	120	60	60	120	240
SEX-2	60	60	120	60	60	120	240
Grand total	120	120	240	120	120	240	480

- Now double click on count of FTOT shown in Fiugre 7.11 and select Average. This gives the avaerage of all scores falling in each combination.
- The contents of Table 7.5 get modified into the new table as shown in Table 7.6.

Table 7.6   Pivot Table of Average FTOT

Average of FTOT	AS = 1		Average 1	AS = 2		Average 2	Grand average
	MGT = 1	MGT = 2		MGT = 1	MGT = 2		
SEX-1	64.90	76.42	70.66	81.95	80.33	81.14	75.90
SEX-2	76.90	63.93	70.42	64.17	71.35	67.76	69.09
Grand average	70.90	70.18	70.54	73.06	75.84	74.45	72.49

- We can also use the true labels like MALE for SEX = 1 and FEMALE for SEX = 2 by using the commands FIND and REPLACE in the EDIT option of the main menu.
- Table 7.6 with all labels appears as shown in Table 7.7.

Table 7.7   Pivot Table with Column and Row Labels

Average of FTOT	AS (Arts)		Total (Arts)	AS (Science)		Average (Science)	Grand average
	MGT (Govt)	MGT (Private)		MGT (Govt)	MGT (Private)		
MALE	64.90	76.42	70.66	81.95	80.33	81.14	75.9
FEMALE	76.90	63.93	70.42	64.17	71.35	67.76	69.08
Grand average	70.90	70.18	70.54	73.06	75.84	74.45	72.49

We can effectively use the pivot table reports to present business data where two-dimensional tables are important. The chief advantage of pivot tables is that once a table is prepared, we can change the summary from one characteristic to another.

7.4 THE CONCEPT OF PROBABILITY

Apart from descriptive statistics, the major function of statistics is to help in drawing conclusions about the population parameters based on sample statistics. This is called *inferential statistics* and an important concept that is related to this branch is *probability*.

'Probability' is basically the chance of happening or non-happening of an event. We use the terms 'perhaps', 'likely', 'possibly', 'may be', and so on, to express our feeling about the uncertainty in the happening of events around us. Consider the following question of a researcher:

1. In my laboratory trials, I have found this drug better to the other. Could this finding be generalized?
2. These two sample observations appear to be related to each other. Is it likely that the same situation prevails with the entire population or in other samples drawn from that population?
3. What is the chance that the call for a blood transfusion from a surgery unit is for a rare group like AB negative?

In all these cases, the uncertainty about the events under question can be expressed by a measure called 'probability' denoted by the letter 'p', lying between 0 and 1. It can also be expressed as the percentage or the proportion of cases observed in favour of the event of interest, out of all the possible cases.

A small value of p indicates a low chance of occurrence while a large value indicates a high chance of occurrence of an event. For instance, if the probability of getting 2 defective floppy disks in a box of 10 is 0.05, it means that only 5% of the boxes are likely to have 2 defective disks. If there is 85% chance of having defect-free boxes, it means the probability of getting zero defectives per box is 0.85. We use this type of statements in *inferential statistics*.

One way of computing the probability is based on the observed frequency table corresponding to a data. If each frequency is divided by the total frequency, we get a number lying between 0 and 1. This is called the *relative frequency* and used as an *estimate* of the probability. In Example 1.2, the relative frequency of having 3 defectives per box is 9/300 which equals 0.03 or 3%.

When we use a variable to explain a random phenomenon, we can explain it in terms of probabilities. Such variables are called *random variables*. The number of TV sets that a shopkeeper could sell in the forthcoming month is something that cannot be predicted with accuracy. So, it is a random variable. The only way of understanding this variable is in terms of probability.

Apart from the empirical probability estimated from a sample data, there are theoretical models that could explain the happening of events. These are called *probability distributions*.

In the next chapter, we use this concept to discuss the inferential aspects of a data.

REFERENCES

1. Microsoft Office—Excel reference from online help.
2. Robert Cowart (1995): *Mastering Windows 95—The Windows 95 Bible*, BPB Publication, New Delhi.
3. Srinivas, P.B. (1999): A study on the academic stress among the students of intermediate classes, part of the doctoral thesis in psychology, S.V. University, Tirupati.

DO IT YOURSELF

7.1 Open the data file TRIBAL.DBF (if not created, do it using Appendix A) and prepare the frequency distribution of the variable EIN and EBAL (see Section 3.2).

7.2 Data need not be entered as a single column. If the data refers to only a single variable, we can enter the values at the rate of 10 per row. While referring to them in the dialogue box for summary statistics, we have to simply block the cells containing the data. Enter the following data in the Excel worksheet and obtain the summary statistics with the help of the Data Analysis Park. The values refer to the mileage obtained by XYZ Company's cars (in kilometres per litre) in an experiment with 30 cars are as under:

19.7	21.5	22.5	22.2	22.5
22.0	22.8	23.2	20.8	21.9
22.5	21.0	19.6	24.1	19.6
23.5	25.3	24.6	25.6	21.3
25.2	19.6	19.5	18.6	24.9
25.6	24.1	18.9	20.6	21.5

Obtain the summary statistics for this data. Also, locate what is the most likely value of the mileage?

7.3 Find different basic statistics for the following few observations by using the Paste functions (and not the Data Analysis Park).

23.2, 23.0, 29.4, 31.0, 31.9, 31.6, 29.3, 28.6, 30.5, 30.7

7.4 Repeat the Problem 7.3 with a simple modification in the data: Subtract 10 from each value and create a new data series. Then find the basic statistics.

7.5 Create a data file with the following 30 records (Table 7.8). The factors are GENDER (Male = 1 and Female = 2), INCOME (Low = 1, High = 2), FTYPE (Family Type—Joint = 1, Nuclear = 2). The response variables are MSS (Marital Satisfaction Score) and STRESS (Stress Score). (Data Courtesy: Vasudha, Department of Home Science, S.V. University, Tirupati.)

Table 7.8

S.No.	Gender	Income	Ftype	MSS	Stress
1	1	1	1	59	7
2	1	2	2	59	5
3	1	1	1	55	9
4	2	1	1	55	8
5	1	1	2	61	7
6	2	2	2	58	8
7	1	2	2	60	5
8	1	2	1	57	7
9	1	1	1	57	6
10	2	2	2	55	9
11	2	1	1	58	7
12	2	2	1	60	5
13	1	2	2	58	9
14	2	2	1	59	7
15	1	2	2	58	7
16	2	1	1	60	6
17	1	2	1	62	6
18	2	1	1	62	8
19	1	1	2	55	7
20	2	1	1	58	6
21	2	1	2	55	8
22	1	2	2	60	6
23	1	1	1	60	7
24	1	1	1	54	8
25	1	1	1	56	7
26	2	1	1	60	4
27	1	2	2	60	5
28	2	1	1	56	7
29	1	1	1	55	7
30	2	1	2	56	9

Use pivot tables and prepare a two-way frequency table showing the number of cases of STRESS according to GENDER (on rows) and FTYPE (on columns). Change the table so as to get the AVERAGE STRESS score according to these factors. Repeat the exercise with MSS.

7.6 The distribution of the number of defects (X) obtained per set in the final inspection of TV sets is given below (Table 7.9):

Table 7.9

No. of defects (X)	No. of cases	No. of defects (X)	No. of cases
0	34	4	14
1	58	5	6
2	54	6	2
3	32		

Enter these values in a worksheet. Find the sum of all cases with the help of the Sum icon. It comes to 200. Now in the next column write the formula +B2/200. Copy the formula to all the other cells in that column. The resulting values are the empirical probabilities corresponding to the values of X. Put a proper heading to this column.

Inferential Statistics Using Excel

Inferential statistics deals with two major activities: (i) estimation of unknown parameters of a population and (ii) testing whether the sample data have sufficient evidence to support or reject a hypothesis about the population parameters. A researcher should be familiar with these two concepts.

8.1 ESTIMATION OF UNKNOWN PARAMETER VALUES

Every population is characterized by certain parameters like mean, median, standard deviation. Each parameter explains one aspect of the population. When we do not have prior knowledge about the value of the population parameter, we use the concept of estimation. An example of a parameter is the mean body mass index (BMI) of male tribals. Since the research data is usually a sample (not the entire population), one way of estimating the mean BMI of the population is to obtain a single value as an *estimate*. It is called a *point estimate*.

A function of the sample observations, like the mean, is called a *statistic*. The statistic used to estimate the parameter is called the *estimator*. A good estimator is characterized by certain properties like *unbiasedness* or *consistency*.

Depending on the sample size and the nature of population, this point estimate may deviate from the true parameter by some amount of unknown and variable error. The relationship between the statistic and the parameter is given by

$$\text{Parameter} = \text{Statistic} \pm \text{Error}$$

If μ denotes the population mean and $\bar{x}$ denotes the sample mean (estimator) then the relationship can be expressed as

$$\mu = \bar{x} \pm \text{error} \quad \text{or} \quad \bar{x} = \mu \pm \text{error}$$

By properly estimating the error, it is possible to provide an interval estimate instead of a point estimate of the population mean. The error is usually specified by the standard deviation of the statistic, which is known as the *standard error*.

While estimating the population mean μ, an interval estimate is of the form

$$\bar{x} \pm 1.96 \frac{\sigma}{\sqrt{n}}$$

where μ is the mean, σ is the standard deviation of the population and n is the sample size. If σ is not known, we use its point estimator in place of σ. The factor 1.96 follows from the theory of normal distribution, according to which, if we construct an interval with approximately two standard deviations (exact value is 1.96) on either side of the sample mean, then there is a 95% chance that the true population mean lies in this interval. This is called *95% confidence interval of μ* given by

$$\bar{x} - \frac{1.96\sigma}{\sqrt{n}}, \qquad \bar{x} + \frac{1.96\sigma}{\sqrt{n}}$$

It is also called the *interval estimate* of the population mean.

Thus, if the sample mean is used as an estimate of the population mean, then in the long run, only 5% of the sample means lie outside the stated interval.

8.2 TESTING OF HYPOTHESES

A statistical hypothesis is a verifiable statement about the population characteristics (parameters) formulated with prior knowledge or based on theoretical considerations.

We use the sample observations to test the likelihood of the truth in the hypothesis. In the light of the sample data, if the hypothesis is found false, we may have evidence to believe in an alternative hypothesis. The following are a few technical terms in this context.

Null hypothesis

It is a hypothesis usually formulated in a way opposite to what we wish to prove. For instance, if we wish to prove that the teaching method A is better than B, we formulate the null hypothesis that there is no *difference between the two methods*. This is usually denoted by H_0. An example of a null hypothesis is

H_0: The average time spent by a professor on computer is 4 hours per day.

The actual research hypothesis tested may be that the average time spent is less than 4 hours a day. In symbols, we write this as

$H_0 = 4$ hours

Alternative hypothesis

The alternative hypothesis is denoted by H_1 and can be in one of the following forms.

$H_1: \mu > 4$ hours

$H_1: \mu \neq 4$ hours (μ not equal to 4 hours)

$H_1: \mu < 4$ hours

It means that whenever the null hypothesis is not true, one of the alternatives must be true. That is why H_0 and H_1 are mutually disjoint statements. In verbal terms, the alternative hypothesis could be like

H_1: The average time spent by a professor on computer is less than 4 hours a day.

It is stated as $H_1: \mu < 4$ hours. This is called the *left-tailed* hypothesis as it addresses the values to the left of the value under the null hypothesis. We can similarly have *right-tailed* hypothesis like $\mu > 4$ hours. Right and left-tailed hypotheses are called *one-sided* or *truncated* hypotheses.

Significance level

Consider a problem of testing, whether the means of two populations really differ or not. Let the following hypothesis be framed.

H_0: The population means are equal ($\mu_1 = \mu_2$)

H_1: The population means are not equal ($\mu_1 \neq \mu_2$)

By looking at the sample means we may arrive at the conclusion that H_0 is wrong and hence support H_1. Suppose H_0 is really true and the sample evidence is a *chance occurrence*, which has led to a wrong rejection of H_0. This is an error, called *Type-I error*. The analyst never wishes to commit this error but it is impossible because the only evidence is a *random sample*. So, it is appropriate to set an *a priori* level like 0.05 or 0.10 as the risk of committing Type-I error. It is called the *level of significance* (LOS) denoted by the Greek letter α. The commonly used values of α are 0.05 or 0.01 which are understood as 5% and 1% levels, respectively.

There is another error that can be committed by an analyst while accepting the null hypothesis even though it is false. It is like accepting that the population means are different even when they do not really differ. This is called the *Type-II error* and the risk of committing this error is denoted by β and it is called the *β-risk*. The mathematical theory of testing of hypothesis is based on the principle that at a fixed level of significance α, the test procedure should have the least β-risk. The quantity $(1 - \beta)$ is called the *power* of the test and it is a measure of the discriminating power of the test between chance occurrence and true occurrence of the test result under the given hypothesis.

Simple hypothesis

A hypothesis is said to be *simple* if it completely specifies the distribution of the population. For instance, in case of normal population with mean μ and standard deviation σ, a simple null hypothesis is of the form $H_0: \mu = \mu_0$, σ known. Since σ is known, knowledge about μ would be enough to understand the entire distribution. For such a test, the probability of committing the type-1 error is expressed as *exactly* α. For instance, if we wish to test the null hypothesis $H_0: \mu = 4$ hours against $H_1: \mu > 4$ hours we say that the test is done with exactly $\alpha = 0.01$, when σ is known.

Composite hypothesis

If the hypothesis does not specify the distribution of the population completely, it is said to be a *composite* hypothesis. Following are some examples:

$H_0: \mu \leq \mu_0$ and σ is known

$H_0: \mu \geq \mu_0$ and σ is known

$H_0: \mu = \mu_0$ and σ is not known

All these are composite because none of them specifies the distribution completely. Hence, for such a test the LOS is specified not as α but as 'at most α'. For instance while testing $H_0: \mu \geq 4$ hours as against $H_1: \mu < 4$ hours, we say that the test has α value at most 0.01. However, while actually carrying the test, we specify the *boundary value* $H_0: \mu = 4$ hours and conduct the test with $\alpha = 0.01$.

Critical value

While testing for the difference between the means of two populations, our concern is whether the observed difference is *too large* to believe that it has occurred just *by chance*. But then the question is *how much* difference should be treated as *too large*? Based on the sampling distribution of the means, it is possible to define a cut-off or threshold value such that if the difference exceeds this value, we say that it is not an occurrence by chance and hence there is sufficient evidence to claim that the means are different. Such a value is called the *critical value* and it is based on the level of significance. For instance, while testing for the significance of the means based on a large sample, the critical value is 1.96 at 5% level of significance for a two-sided test and 1.68 for a one-sided test. The details can be seen from the table of standard normal distribution given in the

Appendix E. The critical value is based on the theoretical distribution of the test statistic under consideration. The following rule would help in reading the critical values from the tables:

If $\alpha = 0.05$ and the test is two-sided, read the critical value under $\alpha = 0.05$.

If $\alpha = 0.05$ and the test is one-sided, read the critical value under $\alpha/2 = 0.025$.

The test statistic

Whenever a test is conducted we accept the null hypothesis if the calculated value of the test statistic is less or equal to the critical value; otherwise it is not accepted. The *test statistic* is a value obtained from a sample data based on a formula. In the conventional way of reporting statistical results, we report (a) the calculated value of the statistic, denoted by Z_{cal} and (b) the critical value denoted by Z_{cri}.

The critical values are available in statistical tables and one has to obtain the same before drawing the inference. If $Z_{cal} > Z_{cri}$, we conclude that "the sample data do not have evidence to accept the null hypotheses and hence it is rejected" at $\alpha\%$ level of significance.

The p-value of a test

The observed significance level is a criterion that can be computed from a sample data. It is the probability of obtaining a test value as large as the observed one, when the null hypothesis was really true. This is called the *p-value* of a test and commonly shown in every computer output associated with a problem of testing. In case of testing the difference between two means, it is the probability of obtaining a difference as large as what has been observed from the sample, when there is really no difference in the population means. Symbolically, the *p*-value is simply $P(Z_{cal} \geq Z_{cri})$. This value is based on the sampling distribution of the Z-statistic. Since it involves complicated mathematical calculations, the *p*-value cannot be worked out manually.

Therefore, the *p*-value is the probability of wrongly rejecting the null hypothesis. If $p < 0.05$, we say that the test result is significant at 5% level. We then reject the null hypothesis with a high confidence. When the *p*-value is given along with the test result, there is no need to specify the critical value.

More details on these aspects can be found in Kachigan (1986), Johnson (1995) or from the Help file in MS-Excel under *testing, probabilities, Z-test*.

8.3 STATISTICAL TESTS CONCERNING MEANS

In this section, we discuss some statistical tests for comparing the means of data groups. These tests are based on the assumption that the sample data has come from a normal population with mean μ and variance σ^2. Knowledge about the population variance is an important factor for analysis. In case of a sample with size 30 or more (usually called a large sample), the sample variance can be used in place of σ^2 and the area property of the normal distribution is used to test the null hypothesis. Such a test is often called the *normal test* or the *Z-test*. For these tests, the test statistic Z follows the standard normal distribution for which the mean is 0 and the standard deviation is 1.

When the sample size is small as in the problems of clinical trials or laboratory tests, the test statistic Z for comparing the means does not follow the standard normal distribution. The exact distribution is called the *t-distribution* after the name of its inventor W.S. Gosset who published his results by the pen name Student. It is popularly known as *Student's t-distribution* and the tests based on it are known as *t-tests*.

The *t*-distribution is indexed by constants called the *degrees of freedom* denoted by DF. The DF is the number of independent observations available for estimating the true parameter of the population. If the sample size is n, then the DF is $(n - 1)$. Depending on the type of test, the DF is appropriately defined. The critical value of the test has to be read from the tables of *t*-distribution corresponding to the DF at the desired level of significance.

Here is one way of understanding the concept of DF. Let us try to write three numbers of our choice. We are free to write any three numbers like 5, 6 and 8. Suppose, we again try to write three numbers but this time under a constraint that their sum should be equal to 20. Then obviously any two numbers can be written independently but not the third number. It means that out of three possible ways of writing, one degree of freedom is lost. Thus every constraint imposed on the data reduces one degree of freedom.

The *t*-distribution possesses all the properties of the standard normal distribution, when the sample size becomes large. In such cases, the critical value for the test can be used either from the *t*-distribution tables or from the table of standard normal distribution. Hence, tests based on the *t*-distribution can be used for large samples without any difference.

Now we discuss some tests for comparison of means.

8.3.1 The One-sample Z-Test for Mean

The objective of this test is to investigate whether the difference between the hypothesised mean (μ_0) and the sample mean ($\bar{x}$) is significant. This test is based on the assumption that the sample data has come from a normal population with mean μ and known variance σ^2. The null hypothesis is $H_0: \mu = \mu_0$ and the alternative can be either two-sided or one-sided.

- When the population variance σ^2 is known and the sample size is large, the test is known as the Z-test and the test statistic is given by

$$Z = \frac{\bar{x} - \mu_0}{\sigma/\sqrt{n}}$$

  For a two-sided test, the critical value at $\alpha = 0.05$ level is 1.96 and for a one-sided test, the critical value at this level is 1.645. The absolute value of Z, obtained by dropping the sign, will be compared with this critical value. If the calculated value Z_{cal} exceeds the critical value, we reject the null hypothesis; otherwise do not reject it.

- When the sample size is large but σ is not known (which is the case usually), the sample standard deviation can be used as an estimate of σ and the same Z-test can be used for testing the significance of the mean.

- When the sample size is small (around 25 or not very less than 30) but σ is known, we can still use the Z-test but it will be an approximate large sample test.

EXAMPLE 8.1 Based on a sample of size 100, the average lifetime of an electronic component is found to be 1,480 hours. The company claims that the average lifetime is 1,500 hours with a standard deviation of 50 hours. Is there sufficient evidence to support the claim of the company?

Analysis Here we have to take the hypothetical mean as $\mu_0 = 1,500$ hours and $\sigma = 50$ hours. Since $n = 100$ cases, we can use the Z-test. Applying the formula given above, we get $Z = -4.00$. Since the absolute value is 4.00 it is greater than 1.96 and we reject the null hypothesis that the sample mean and the hypothetical mean do not differ significantly. It means that the average life cannot be 1,500 hours and hence there is no evidence to support the claim of the company. ■

Excel however does not have a separate module to carry out this test, when only mean and standard deviation are given.

8.3.2 The One-sample *t*-Test for Mean

The *t*-test is one of the popularly used tests for comparing the means.

When the sample size is small and the population variance σ^2 is not known, the Z-test cannot be applied for testing the significance of the observed mean. However, when the sample has come from a normal distribution (at least with close approximation) we can use the theory of Student's t-distribution in place of the normal distribution used in the Z-test. The resulting test is called the t-test.

The t-test procedure for testing the significance of the difference between the hypothetical mean and the sample mean is similar to the Z-test. The only difference is that the sample standard deviation (refer to Section 1.9.3) is used in place of σ and the critical value is determined from the tables of t-distribution corresponding to $(n - 1)$ degrees of freedom. The test statistic is given by

$$t = \frac{\bar{x} - \mu_0}{s/\sqrt{n}}$$

and it follows t-distribution with $(n - 1)$ degrees of freedom. It can be both negative as well as positive and we consider the absolute value only.

The critical value is found from the tables of t-distribution at $\alpha/2$ for two-sided test and at α for the one-sided test. The values of t-distribution are given in Appendix-D.

Consider Example 8.2 for the application of the t-test.

EXAMPLE 8.2 The systolic blood pressure (Hg/mm) of 15 patients under treatment for hypertension has been noted as shown below:

150 97 107 139 118 157 152 138 151 133 165 150 122 188 151

Is there evidence to believe that this sample represents a population of patients having a normally distributed blood pressure with mean above 150?

Analysis Here the data is a small sample and the population variance is not known. Then we have to use t-test. We take the null hypothesis as H_0: $\mu = 150$ and consider the one-sided alternative H_1: $\mu > 150$. Let us take the sample data to the *Excel* worksheet. We can find the mean and standard deviation of these 15 values with the help of the Paste function. This gives $\bar{x} = 141.2$ and $s = 23$. The value of the test statistic becomes $t = -1.4819$. The degrees of freedom are $(n - 1) = 14$. If the level of significance is taken as $\alpha = 0.01$, this one-sided test will have a critical value of 1.761. Since the absolute value of the t-statistic is smaller than the critical value, we cannot reject the null hypothesis. It means that the sample is likely to have come from the hypothetical population. ∎

Excel does not provide a module to perform this one-sample t-test directly, with only mean and standard deviation.

8.3.3 The Two-sample Z-Test for Means

This test is used for comparing the means of two independent populations. Let the means of these two populations be μ_1, μ_2 and variances σ_1^2, σ_2^2 respectively. We wish to test whether the difference between these two means could be taken as zero (or some other constant). Let us take two random samples of size n_1 and n_2 respectively from these two populations. The test depends on the difference between the sample means $\bar{x}_1$ and $\bar{x}_2$. The objective is to investigate whether the observed difference between the sample means is any evidence of a significant difference between the two population means. The test statistic is given by the expression

$$Z = \frac{\bar{X}_1 - \bar{X}_2}{\sqrt{\frac{\sigma_1^2}{n_1} + \frac{\sigma_2^2}{n_2}}}$$

and it follows standard normal distribution. As usual, we reject the null hypothesis if the calculated value of Z is larger than the critical value at the fixed level α. We use Z-test when the sample size is large and the population variances are known. Excel has a tool in the data analysis park to perform this test.

Let us examine the application of this procedure with the help of Example 8.3:

EXAMPLE 8.3 Consider the file C:\STATMAN\TRIBAL.DBF. We wish to compare the means of BMI corresponding to the tribes Sugali and Yanadi. In the data file, we have designated them as code = 1 and code = 2, respectively.

We wish to test whether the difference in the mean BMI of the two groups is statistically significant. The population variances are known to be 5 and 7 respectively for the two groups.

Analysis To answer this question, we first open the data file. Then select the Data Analysis Park. The last item in the menu is the Z-test for two sample means. Select it and click OK. We get the dialogue box as shown in Figure 8.1. Since we are using an already existing file with all records we have to pick up the required records and input them. We can proceed as follows to do this job.

- Open the data file in Excel
- Select the Z-test for two sample means in the Data Analysis Park
- Click on the window corresponding to Variable-1 Range
- Select with mouse all cells under the heading BMI corresponding to Code = 1
- For the Variable-2 Range, select with mouse all cells under the heading BMI corresponding to Code = 2

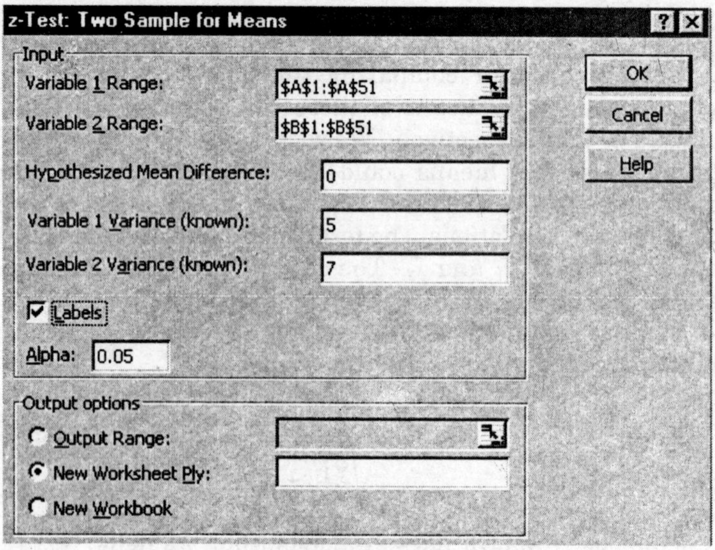

Fig. 8.1 Dialogue box for the two-sample Z-test.

- Type 0 against Hypothesised Difference
- Type 5 and 7 in the windows corresponding to the variances of variables 1 and 2 respectively. These are hypothesised variances of the population. The reader can change these values and observe the result
- Click on Labels
- Click OK.

Now the Excel output appears as shown below.

Z-Test: Two-sample for Means

	Variable 1	Variable 2
Mean	21.0156	18.96
Known variance	5	7
Observations	50	50
Hypothesised mean difference	0	
Z	4.19598	
$P(Z <= z)$ one-tail	0.00001	
z Critical one-tail	1.64485	
$P(Z <= z)$ two-tail	0.00003	
z Critical two-tail	1.95996	

From the test results it follows that the calculated value of Z is 4.19598 which is larger than the critical value at the 5% level. If we wish to change this level we have to enter 0.01 in the dialogue box against the Alpha. By default this value is taken as 0.05. ∎

This Z-test can be applied only when the data is large and the population variances are known. When they are not known we use the sample variances in place of population variances. Excel does not provide a test of this type. Instead, it provides a t-test that can be used even with small samples. When the sample size is large this t-test automatically becomes the Z-test. So, we can safely use t-test for comparing two means irrespective of whether the sample size is large or small.

8.3.4 The Two-sample *t*-Test for Means

This is a test for comparing the means of two populations based on the sample means.

This is commonly known as the small sample test for means. The test procedure is similar to that of the Z-test except that the critical value is based on the t-distribution instead of the normal distribution. Here is an example.

EXAMPLE 8.4 Consider the following data on the BMI of tribals corresponding to the two tribes Sugali and Yanadi collected from the file TRIBAL.

Sugali	Yanadi	Sugali	Yanadi
20.43	17.70	18.08	18.60
22.51	21.40	20.63	18.50
18.99	20.70	22.55	18.20
20.49	19.30	22.43	20.30
23.12	21.00	22.77	
25.63	17.90	23.23	

Let us use the t-test for comparing the means of these two groups. Note that the sample sizes are not the same.

Analysis While proceeding with t-test the Data Analysis Park offers two types of t-tests. The first one is used when the researcher has belief (evidence) that the population variances are equal. The other one is used when the population variances are likely to be different. To perform the t-test we proceed as follows.

- Enter the data in a worksheet.
- Select the t-test for equal variances case from the Data Analysis Park.
- Select the data cells having the input for the variables Sugali and Yanadi.
- Set the hypothesised difference as zero.

The output of the t-test appears as follows.

t-Test: Two-sample Assuming Equal Variances

	Sugali	Yanadi
Mean	21.7383	19.36
Variance	4.3192	1.8982
Observations	12	10
Pooled variance	3.22977	
Hypothesised mean difference	0	
Df	20	
t-Stat	3.09077	
$P(T <= t)$ one-tail	0.00288	
t Critical one-tail	1.72472	
$P(T <= t)$ two-tail	0.00576	
t Critical two-tail	2.08596	

For the same data if we select the option Unequal Variances, we get the following output.

t-Test: Two-sample Assuming Unequal Variances

	Sugali	Yanadi
Mean	21.7383	19.36
Variance	4.3192	1.8982
Observations	12	10
Hypothesised Mean Difference	0	
Df	19	
t-Stat	3.2077	
$P(T <= t)$ one-tail	0.0023	
t Critical one-tail	1.7291	
$P(T <= t)$ two-tail	0.0046	
t Critical two-tail	2.0930	

From the results of these two t-tests, we can observe that the difference between the sample means is significant and recommends to reject the null hypothesis at $\alpha = 0.05$. ■

The case of equal variances is normally used in the analysis. However, if the researcher has knowledge about the population variances being not equal, then the latter procedure can be followed. It is known as *Smith–Satterthwaite test*, which is used in place of the usual two-sample t-test. More details about various applications of t-test can be found in Johnson (1995).

The tests conducted with the help of the Data Analysis Pak can also be performed with the help of the Paste functions. For instance, by selecting the item TTEST from the Statistical functions of the

Insert function menu we can carry out this test directly. We then get the following dialogue box in which we enter the data ranges. Array1 refers to the first data set and Array2 refers to the second data set. If we wish to use one-tailed test we type 1 and for a two-tailed test we type 2 in the window against Tails. The type of the test should also be entered by the corresponding code as shown in Figure 8.2. Even though we can use this function to conduct the t-test, the

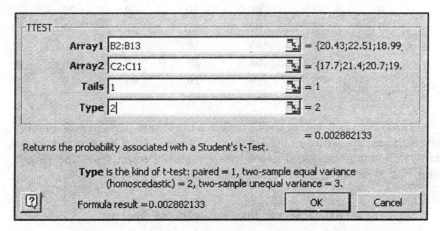

Fig. 8.2 TTEST Paste function and the output.

output given by the Data Analysis Park contains more details. So this can be used only as a quick test. The result is indicated by the p-value corresponding to the two-tailed t-test. Since this value is smaller than 0.01, we can consider the difference between the means as significant.

8.3.5 The Paired t-Test

Sometimes the data in the two samples is in the form of pairs instead of random groups. It is like taking a clinical measurement before and after a treatment. In such cases, the sample data will be correlated to each other and the t-test to be used is called the *paired* t-test. The procedure is based on the difference between the two groups of paired data. The mean difference is then tested against a hypothetical value 0. The null hypothesis is that there is no difference between the two means.

Here is an example for illustration.

EXAMPLE 8.5 The following data refers to the daily earnings of 20 individuals after a training program under self-employment.

Table 8.1 Daily Earning of 20 Individuals

S.No.	Before	After	S.No.	Before	After
1	80	90	11	70	75
2	75	70	12	80	85
3	55	60	13	85	82
4	60	60	14	90	87
5	65	68	15	75	72
6	55	62	16	70	72
7	65	60	17	65	65
8	60	65	18	60	50
9	70	65	19	50	55
10	75	80	20	65	70

Analysis Let us test the hypothesis that there is no improvement after giving the training. Then the hypothesised difference is taken as 0. Following the usual procedure for paired t-test from the Data Analysis Pak, we get the following output.

t-Test: Paired Two-sample for Means

	Before	After
Mean	68.5	69.65
Variance	110.7895	118.7658
Observations	20	20
Pearson's Correlation	0.8830	
Hypothesised Mean Difference	0	
Df	19	
t-Stat	-0.9902	
$P(T \leq t)$ one-tail	0.1673	
t Critical one-tail	1.7291	
$P(T \leq t)$ two-tail	0.3345	
t Critical two-tail	2.0930	

The t-value (after dropping the negative sign) is found to be very small when compared with the one-sided critical value. It shows that the difference is not significant.

Suppose we make the hypothesised difference as 10 instead of 0. We get a totally different result as shown below.

t-Test: Paired Two-sample for Means

	Before	After
Mean	68.5	69.65
Variance	110.7895	118.7658
Observations	20	20
Pearson's Correlation	0.88302	
Hypothesised Mean Difference	10	
Df	19	
t-Stat	−9.60060	
$P(T <= t)$ one-tail	0.00000	
t Critical one-tail	1.72913	
$P(T <= t)$ two-tail	0.00000	
t Critical two-tail	2.09302	

From the null hypothesis, the average difference is 10. Such a strong assertion could not be accepted since the calculated t-value is higher than the critical value. Further, the p-value is so small to believe that the difference has occurred due to chance. This shows that the results get very much altered when the hypothesised difference is chosen as something other than zero. ∎

This paired t-test can also be conducted from the Paste functions.

8.4 THE F-TEST FOR VARIANCE

While using the t-test for comparing two means, we have assumed that the variances of the two populations are equal. This assumption can be verified by testing the hypothesis $H_0: \sigma_1^2 = \sigma_2^2$. This test is called the *variance ratio* test. Excel has a module in the Data Analysis Park to do this job. If n_1, n_2 denote the sizes of the two samples respectively and S_1^2, S_2^2 denote the corresponding estimates of the population variances based on the two samples, then the F-statistic is given by $F = S_1^2/S_2^2$.

This statistic follows a distribution called the Snedecor's F-distribution with $(n_1 - 1, n_2 - 2)$ DF. It is assumed that the populations from which the two samples have been drawn are normally distributed.

For the one-sided alternative hypothesis $H_1: \sigma_1^2 > \sigma_2^2$, the critical value is denoted by F_α. We reject H_0 and accept H_1, if the calculated F value (F_{cal}) exceeds F_α.

If the alternative is $H_1: \sigma_1^2 < \sigma_2^2$, the critical value is $F_{1-\alpha}$ with $(n_2 - 1, n_1 - 1)$ DF. We accept H_1 if $F_{cal} < F_{1-\alpha}$. The table of critical values are usually given at $\alpha = 0.05$ and $\alpha = 0.01$ levels. The DF of the numerator in F is shown on the top of the table and the DF of the denominator is shown along the rows. As a matter of convenience, while conducting this test, we keep the larger variance in the numerator so that the calculated value is always ≥1.

The Excel output takes care of the results to be presented. Here is an example.

EXAMPLE 8.6 Consider the data given in Example 8.2. We wish to test the null hypothesis that the population variances of the two tribal groups do not differ significantly.

Analysis To do this, we choose the option F-test for Two Variances from the Data Analysis Pak. Proceeding as in the case of the other tests discussed earlier, we get the following output.

F-Test: Two-sample for Variances

	Sugali	Yanadi
Mean	21.7383	19.36
Variance	4.3192	1.8982
Observations	12	10
Df	11	9
F	2.2754	
$P(F <= f)$ one-tail	0.1136	
F Critical one-tail	3.1025	

It follows from the table that the F-ratio is smaller than the critical value and hence we do not reject the null hypothesis. Hence the population variances can be taken as equal. Such populations are known as *homoscedastic*. ■

8.5 ANALYSIS OF VARIANCE

This is a statistical test for comparing the means of more than two populations or groups. With two means, we use t-test or the Z-test but when there are three or more groups for comparison, we have to use the procedure called analysis of variance or simply ANOVA. It gives an overall comparison of means and investigates whether the population means are likely to be same or different. If they are found to be different then there are other tests to verify which pair of means has a significant difference.

Excel has a module in the Data Analysis Pak to perform ANOVA. It appears in three forms: (a) one-way ANOVA (b) two-way ANOVA with replication and (c) two-way ANOVA without replication.

- When the data is classified into groups according to only one characteristic or factor, it is called *one-way classification* and the corresponding ANOVA is called the one-way ANOVA or *one-factor* ANOVA. Examples are groups of individuals classified according to gender, tribe, caste, etc.

- When the data is classified according to two characteristics or factors, like age group and gender we call it a two-way classified data and the corresponding ANOVA is called the *two-way* ANOVA *or two-factor* ANOVA. At each combination of the levels of the two factors, there may be more than one data value. This is called *replication*. When there are replications, it is possible to estimate the *interaction* or the joint effect of the two factors on the response being studied.
- When there are no replications, we can still perform two-way ANOVA. In this case, interactions cannot be estimated. The Excel procedure provides separate analysis for this type of data.
- The null hypothesis in ANOVA is that where the group means are equal. The method is based on partitioning the total variation present in the data into variation due to (a) factors and (b) due to error (or non-specific reasons). The test criterion is the F-ratio and the null hypothesis is rejected whenever the calculated F-ratio exceeds the critical value.
- The results of ANOVA are given in the form of a table called the ANOVA *table*.

Let us see these aspects with some examples and Excel procedures.

8.5.1 One-way ANOVA

In this section, we discuss a case of one-way ANOVA with the help of data obtained from the file TRIBAL. The details of the data and the usage of the tool are discussed below.

EXAMPLE 8.7 The following data gives the BMI values of the tribal men belonging to three groups. This is a one-way classified data since the data values are grouped according to Tribe only and the data looks as shown below.

Sugali	Yanadi	Yerukala
20.43	17.70	25.10
22.51	21.40	21.80
18.99	20.70	19.80
20.49	19.30	18.40
23.12	21.00	18.80
25.63	17.90	20.70
18.08	18.60	18.80
20.63	18.50	19.40
22.55	18.20	18.50
22.43	20.30	18.40

We can now compare the means of these three groups and test the following null hypothesis using ANOVA.

H_0: The population means of the three groups do not differ significantly

The alternative hypothesis is

H_1: At least one pair of population means has a significant difference.

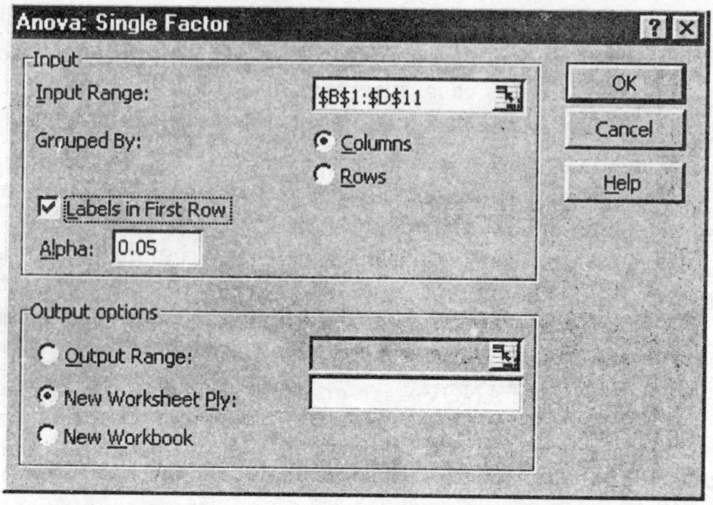

Fig. 8.3 Dialogue box for single-factor ANOVA.

Analysis To carryout ANOVA, we select the module ANOVA: Single Factor from the Data Analysis Pak and click the corresponding input cells. The dialogue box appears as shown in Figure 8.3. It is assumed that the data has come from a normally distributed population and that the variances in different groups are equal. With all the entries filled in the dialogue box, the output appears as shown below.

ANOVA: Single Factor

SUMMARY

Groups	Count	Sum	Average	Variance
Sugali	10	214.86	21.486	4.8428
Yanadi	10	193.60	19.360	1.8982
Yerukala	10	199.70	19.970	4.4868

ANOVA

Source of variation	SS	DF	MS	F	p-value	F crit
Between Groups	23.9674	2	11.9837	3.2020	0.0565	3.3541
Within Groups	101.0502	27	3.7426			
Total	125.0176	29				

There are two major units in the output. The first part gives the summary statistics of the variable BMI for all the groups. The purpose is to provide the estimates of the population means. The second part is the ANOVA, which deals with testing the null hypothesis

H_0: The mean BMI of Sugali, Yanadi and Yerukala are the same

against the alternative hypothesis

H_1: The three means are not the same.

In the ANOVA table, the total variation is split into two parts and shown as 'between groups' and 'within groups'. Since there are 3 groups, the DF is 2. The DF for the item Total is $(30 - 1) = 29$. The term SS stands for Sum of Squares and the term MS stands for Mean Sum of Squares. The MS is an estimate of the variation in the data contributed by the groups. The remaining variation is attributed to 'within groups'. It is also called the *residual* or *error*.

$$\text{The } F\text{-ratio is defined as } F = \frac{\text{MS between groups}}{\text{MS within groups}}$$

If the groups really have any significant effect, we can expect the F-ratio to be larger than the critical value. (If the F-ratio is less than 1, it is automatically insignificant and we need not compare it with the critical value.) Now let us understand the output given by Excel.

The results of the ANOVA show that the variation between the tribal groups is *not significant* because the F value is smaller than the F-critical value shown in the last column of the table. Further, the p-value is larger than 0.05 and hence the tribe factor has no significant effect on BMI. The null hypothesis is therefore not rejected and we conclude that the mean BMI could be the same for all the three tribal populations. ∎

Remark: The procedure for ANOVA is based on the assumption that the each column of the worksheet represents one group of data values. If different groups have been entered in the same column (as we did in the main TRIBAL.DBF file), Excel cannot handle such data. So, we have to either enter the data as per the classification or pick up the relevant columns from the data file and place them in adjacent columns.

Here is another example with unequal number of values in each group.

EXAMPLE 8.8 Neeharika Pickles is a popular pickle industry dealing with pickles of different types. The management feels that the type of packaging has an influence on the sales. The packing should be both leak-proof and also create a motivational effect on the customers. The company has tried four different types of packaging methods denoted by the codes P, Q, R and S. Data on the number of packets sold during the days of week has been collected and recorded as follows.

Day	P	Q	R	S
Mon	35	37	25	34
Tue	38	35	22	31
Wed	31	35	24	29
Thu	40	34	29	28
Fri		40	31	30
Sat		38	26	

The sales corresponding to methods P and S for Friday and Saturday were not available for analysis.

Analysis Let us compare the average sales through one-way ANOVA. We enter these values in a worksheet and proceed in the same way as done in the previous exercise. We get the following output.

ANOVA: Single Factor

SUMMARY

Groups	Count	Sum	Average	Variance
P	4	144	36	15.33333
Q	6	219	36.5	5.1
R	6	157	26.16667	10.96667
S	5	152	30.4	5.3

ANOVA

Source of variation	SS	DF	MS	F	p-value	F Crit
Between groups	402.4667	3	134.1556	15.4585	4.15E-05	3.196774
Within groups	147.5333	17	8.678431			
Total	550	20				

From the F-ratio, it is clear that the group means do not differ significantly from each. In other words, packing style has no influence on the sales. ■

In this problem, the number of observations is not the same in each column. While entering the data, we have to leave a blank when there is no data. If we put any other symbol like '–' or NA (to mean not available), Excel shows error saying that there is a non-numeric data. If we leave a cell blank then Excel will omit such cells and computes the statistics only on non-empty cells. This is one chief advantage of using Excel for data entry.

The interpretation of the results is on similar lines to that of Example 8.7.

With only two groups, the one-way ANOVA and the Student's t-test will give the same decision. It can be shown that the square of the t-statistic follows F distribution. Hence, we can use either t-test or ANOVA for comparing two means.

8.5.2 Two-way ANOVA with Replication

Let us now consider the ANOVA with two factors with replications. This type of situation usually arises in laboratory experiments and clinical trials where the entire experimental data is classified according to two factors and at each combination of the factors, more than one observation (called response) is obtained from the experiment.

There will be three null hypotheses, one for each of the two factors and another one regarding the interaction. These are stated as follows.

H_{01}: The means of the first factor do not differ significantly (row factor).

H_{02}: The means of the second factor do not differ significantly (column factor).

H_{03}: The interaction between the two factors is not significant.

Let us look into this type of data with the following example.

EXAMPLE 8.9 As part of testing four different types of teaching methods, a researcher has selected three students from three schools located in rural, town and city areas. Thus, the location of the school is a factor and teaching method is another factor. This experiment requires $4 \times 3 \times 3 = 36$ students classified according to *location* and the *teaching method*. In an aptitude test, the score out of 100, is the response obtained from each student as shown in Table 8.2).

Table 8.2 The Score of Students in an Aptitude Test

Method	Rural	Town	City
A	55	58	63
A	59	60	65
A	58	57	68
B	61	65	69
B	61	66	69
B	60	63	70
C	59	60	71
C	60	68	70
C	62	63	69

We wish to test the null hypotheses (i) there is no effect of teaching method on the marks (ii) there is no effect of location of school on the marks and (iii) there is no interaction effect of teaching method and location of school on marks.

Analysis Two-factor ANOVA is the technique to be used for the analysis of this problem. To do this, we first enter the data in Excel with treatments along the rows and the school location in the columns. From the Data Analysis Pak, we have to select the item Two-factor ANOVA with Replication. Then we get the dialogue box as shown in Figure 8.4.

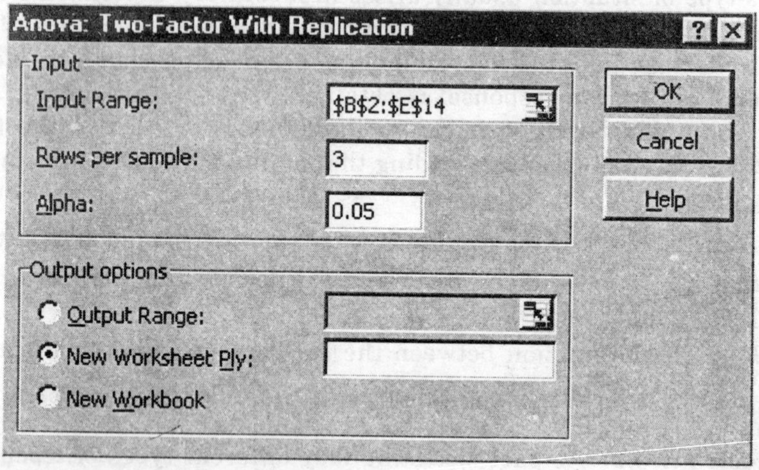

Fig. 8.4 Dialogue box for two-way ANOVA.

We have to enter the data range, the number of rows under each treatment (called rows per sample) and then click OK.

The output of the ANOVA given by Excel appears in the form of several tables each pertaining to the summary statistics of factors along with the ANOVA table. The output is as shown below.

ANOVA: Two-factor with Replication				
Summary	Rural	Town	City	Total
		A		
Count	3	3	3	9
Sum	172	175	196	543
Average	57.333	58.333	65.333	60.333
Variance	4.333	2.3333	6.3333	17.5
		B		
Count	3	3	3	9
Sum	182	194	208	584
Average	60.667	64.666	69.333	64.888
Variance	0.3333	2.3333	0.3333	14.861

	C			
Count	3	3	3	9
Sum	181	191	210	582
Average	60.333	63.666	70	64.667
Variance	2.3333	16.333	1	23
	D			
Count	3	3	3	9
Sum	183	190	215	588
Average	61	63.333	71.666	65.333
Variance	4	14.333	2.3333	28.75
	Total			
Count	12	12	12	
Sum	718	750	829	
Average	59.833	62.5	69.083	
Variance	4.3333	13	7.7196	

ANOVA

Source of variation	SS	DF	MS	F	p-value	F crit
Sample	146.75	3	48.9167	10.4201	0.0001	3.0088
Columns	544.0556	2	272.0278	57.9467	0.0000	3.4028
Interaction	16.1667	6	2.6944	0.5740	0.7471	2.5082
Within	112.6667	24	4.6944			
Total	819.6389	35				

From the summary tables we can read the mean and the variance of the marks corresponding to the four treatments A, B, C and D. The last table having the title *Total* gives the mean and variance of all the 12 observations under each column for Rural, Town and City students. The column headings are common to all the tables.

Let us examine the ANOVA table. The first source of variation is given as Sample, which corresponds to Teaching Methods (rows) and the second source corresponds to Location (columns). Since there is no provision for Labels to be highlighted in the dialogue box, the results of ANOVA do not contain the original headings. The F-values corresponding to Methods and Location are higher than the corresponding critical values and hence we conclude that the both teaching method and the location of school have significant effect on the average marks of the student.

The third source of variation is the Interaction, which is an indication of the joint effect of teaching methods and the location of school, on marks. The effect is however not significant because the F-value obtained is lower than the critical value.

The last term is as usual. The variation within the samples is called the *experimental error* or *residue*. ∎

Now we consider the two-way ANOVA without replication.

8.5.3 Two-way ANOVA without Replication

This technique is applied when the analyst obtains only one response value at each combination of the two factors. In the previous example, suppose only one student is tested under each method from each school. We then get only 12 students in the analysis. In this case, we cannot estimate the interaction effect. Consider the following example.

EXAMPLE 8.10 One student from each location of school is selected and trained under the four teaching methods. The marks obtained in the test are as under (Table 8.3):

Table 8.3 Mark Obtained under Four Teaching Methods

Method	Rural	Town	City
A	55	58	63
B	61	65	69
C	59	60	71
D	59	59	72

We wish to test whether the average marks differ with (a) teaching method and (b) location of school.

Analysis By selecting the module ANOVA: TWO-FACTOR WITHOUT REPLICATION from the Data Analysis Pak and proceeding as done in the earlier example, we get the following output.

ANOVA: Two-factor without Replication

Summary	Count	Sum	Average	Variance
A	3	176	58.667	16.333
B	3	195	65	16
C	3	190	63.333	44.333
D	3	190	63.333	56.333
RURAL	4	234	58.500	6.333
TOWN	4	242	60.500	9.667
CITY	4	275	68.75	16.25

ANOVA

Source of variation	SS	DF	MS	F	p-value	F crit
Rows	66.917	3	22.306	4.486	0.056	4.757
Columns	236.167	2	118.083	23.749	0.001	5.143
Error	29.833	6	4.972			
Total	332.917	11				

The output is self-explanatory. The columns (location of school) have a significant effect because the F-value is larger than the critical value. The methods however do not differ significantly. ∎

ANOVA is thus an important inferential tool in statistics. In many real applications, a two-factor ANOVA would be sufficient because the analyst may not be interested in higher order interactions in a single experiment or survey. Special statistical software is available to carry out ANOVA with more than two factors.

In the following section, we examine a procedure for comparing the frequencies and testing for discrepancy between the observation and belief, as evidenced by sample data. This is done with the help of a test called the *chi-square test*.

8.6 THE CHI-SQUARE TEST

The chi-square test is a statistical test used to compare the observed frequencies with those, which are expected according to some theory or a hypothesis. There are two specific applications of the chi-square test namely (a) chi-square test for the goodness of fit and (b) chi-square test for the independence of attributes. Let us discuss these two areas in detail.

8.6.1 Chi-square Test for Goodness of Fit

This test is used to compare the differences between the observed frequencies and the expected frequencies corresponding to n categories or classes. These expected frequencies could be obtained by using a statistical distribution like binomial, Poisson, normal or some other law.

Let $O_1, O_2, ..., O_n$ be the observed frequencies corresponding to the n classes of data. The theoretical frequencies of these classes may be denoted by $E_1, E_2, ..., E_n$ respectively.

Now the problem is to verify whether the theory according to which these frequencies have been obtained fits well to the data. In other words, we may investigate whether the available data supports the theory or not. The procedure is to compare the O_i and the E_i values and investigate whether the discrepancy could be taken as zero. The null hypothesis is that the theory fits well to the data and the alternative is that the fit is not good. The test criterion is

$$\chi^2 = \sum_{i=1}^{n} \frac{(O_i - E_i)^2}{E_i}$$

where O_i and E_i represent the observed and the expected frequencies of the ith class. This test contains $(n - 1)$ degrees of freedom (DF).

If the calculated value of χ^2 exceeds the critical value for this DF at the chosen level of significance α, we consider the fit to be not good. Here is an example.

EXAMPLE 8.11 The number of out patients visiting the senior medical officer during the weekdays at the University Health Centre has been studied during a week and the following data is obtained (Table 8.4).

Table 8.4 The Number of Out Patients Visiting in a Week

Day	Number of patients
Monday	35
Tuesday	41
Wednesday	46
Thursday	39
Friday	31
Saturday	42
Total	234

We wish to test the null hypothesis that the number of patients is equally distributed during the weekdays.

Analysis To test this hypothesis, we have to work out the expected frequencies assuming the null hypothesis to be true. Then on every day we expect (234/6) = 39 cases. The method is as follows:

- Create a worksheet file and enter the observed and expected frequencies.
- Click on the Paste functions and select CHITEST from it.
- A dialogue box appears on the sheet.
- After entering the array ranges for the observed and expected values we get the chi-square value displayed on the same chart. Each item on the box is explained for online help.
- The data and the result are shown in Figure 8.5.

The value of CHITEST is therefore 0.6021. It is the *p*-value of the test with $n - 1 = 5$ degrees of freedom. In order to set the χ^2-valve we select the PASTE FUNCTIONS and use the CHIINV function from them. It gives a dialogue box as shown in Figure 8.6 in which we enter the calculated value of CHITEST and the degrees of freedom. Then Excel gives the chisquare corresponding to the test as 3.641. We conclude that the fit is good only if the *p*-value is less than the target of $\alpha = 0.05$ or 0.01. The result is shown in Figure 8.6. Since the *p*-value is higher than 0.05 we accept the null hypothesis and conclude that the number of patients are equally distributed during the weekdays. ∎

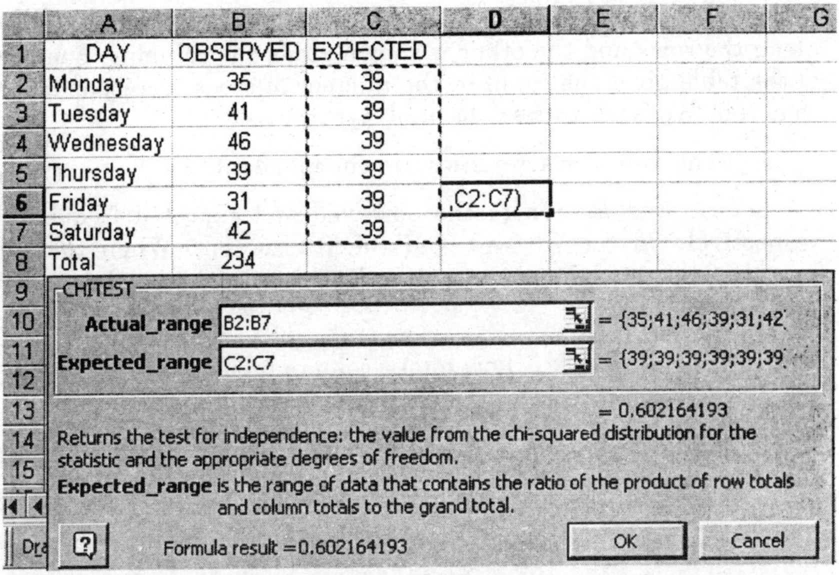

INFERENTIAL STATISTICS USING EXCEL

	A	B	C	D	E	F	G
1	DAY	OBSERVED	EXPECTED				
2	Monday	35	39				
3	Tuesday	41	39				
4	Wednesday	46	39				
5	Thursday	39	39				
6	Friday	31	39	C2:C7			
7	Saturday	42	39				
8	Total	234					

CHITEST

Actual_range B2:B7   = {35;41;46;39;31;42}

Expected_range C2:C7   = {39;39;39;39;39;39}

= 0.602164193

Returns the test for independence: the value from the chi-squared distribution for the statistic and the appropriate degrees of freedom.

Expected_range is the range of data that contains the ratio of the product of row totals and column totals to the grand total.

Formula result = 0.602164193   OK   Cancel

Fig. 8.5 Worksheet function used for chi-square test.

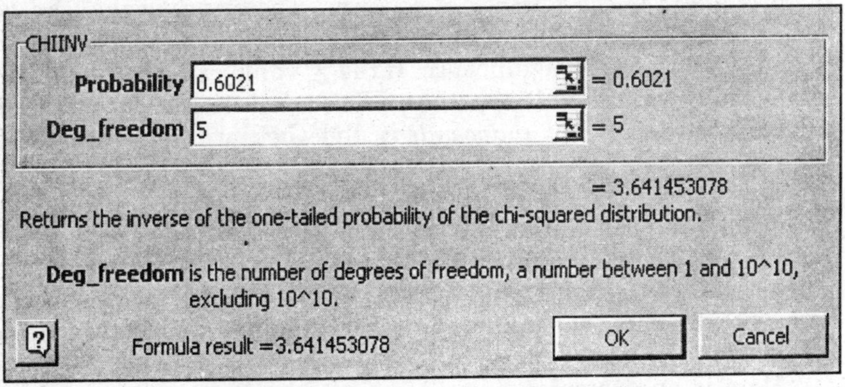

CHIINV

Probability 0.6021   = 0.6021

Deg_freedom 5   = 5

= 3.641453078

Returns the inverse of the one-tailed probability of the chi-squared distribution.

Deg_freedom is the number of degrees of freedom, a number between 1 and 10^10, excluding 10^10.

Formula result = 3.641453078   OK   Cancel

Fig. 8.6 Calculation of chi-square value.

Thus the chi-square test can be used to test for the discrepancy between the observed and expected number of cases in a data.

Let us now consider another application of chi-square test called the *test for independence of attributes*.

8.6.2 Chi-square Test for Independence

Consider the cross-tabulation of some characteristic across two categorical variables. The resulting table is called a *two-way* frequency table or a *contingency* table. One characteristic or attribute is shown along the rows and the other is shown along the columns. Each cell of the table gives the count or the number of cases corresponding to that cell. We wish to test the null hypothesis

H_0: The two characteristics are independent.

Let O_{ij} and E_{ij} denote the observed and expected frequencies respectively in the ith row and the jth column. When the null hypothesis is true, the expected frequencies are calculated according to the formula

$$E_{ij} = \frac{\text{Row total} \times \text{column total}}{\text{Grand total}}$$

After finding the expected frequencies of all the cells, we calculate the chi-square value given by

$$\chi^2 = \sum_{i=1}^{r} \sum_{j=1}^{c} \frac{(O_{ij} - E_{ij})^2}{E_{ij}}$$

The degrees of freedom is $(r-1) \times (c-1)$, where r is the number of rows and c is the number of columns. The critical value can be seen from statistical tables corresponding to these degrees of freedom at the chosen level of significance. If the χ^2 value exceeds the critical value, we reject the null hypothesis. The conclusion is that the two characteristics are *not independent* and they are *associated with each other*.

Excel does not have a direct analysis tool to perform this test of independence. The Chi-test paste function can however be used to do this job. The only difficulty is that Excel does not automatically calculate the expected frequencies for the contingency table. We have to find them, enter them in separate cells and then show these cells as input to Chi-test.

Here is an example.

EXAMPLE 8.12 Table 8.5 gives the distribution of students according to the family type and the anxiety level.

Table 8.5 Distribution of Students

Family type	Anxiety level		
	Low	Normal	High
Joint family	35	42	61
Nuclear family	48	51	63

We wish to test whether the anxiety level and the type of family are independent.

Analysis This is a contingency table with 2 rows and 3 columns. Let us first enter these observed frequencies in the cells of the Excel worksheet. We have to find out the expected frequencies to each cell and then get the chi-square value. The expected frequencies in the cells are shown below.

Family type	Anxiety level		
	Low	Normal	High
Joint family	38.18	42.78	57.04
Nuclear family	44.82	50.22	66.96

Entering these values in the Excel worksheet and using the CHITEST function followed by CHIINV we get chi-square value as 1.025 and the degrees of freedom is 2. The critical value from tables at 0.05 level of significance is found to be 5.991. Since the calculated value of chi-square is smaller than the critical value, we consider the anxiety level and family type as independent. ■

Now these calculations can be simplified with the help of a FoxPro program given in Chapter 10. The program first asks for the number of rows and the number of columns. The individual values have to be entered for each row separately. After all the rows are entered, we get the output showing the actual data and the chi-square value. We can compare this value with the critical value.

We close this chapter with these tests. In the next chapter we proceed with some simple FoxPro programs that can be loaded into the user's computer for performing statistical calculations.

REFERENCES

1. Kachigan Sam Kash (1986): *Statistical Analysis: An Interdisciplinary Introduction to Univariate and Multivariate Methods*, Radius Press, New York.
2. Microsoft Office—Excel reference from online help.
3. Richard A. Johnson (1995): *Miller & Freund's Probability and Statistics for Engineers*, 5th ed., Prentice-Hall of India, New Delhi.

DO IT YOURSELF

8.1 Use the data file TRIBAL and select the cases of EBAL for SEX = 1 by filtering the column under SEX. Now copy these cases to another worksheet and paste them in one column. Again go to the previous sheet and select the cases for SEX = 2. Paste these cells in the next worksheet adjacent to the previously pasted column. Now your data will have two columns of EBAL, one for males and another for females. Carry out t-test for the means of EBAL of these two groups.

8.2 The marital satisfaction score (MSS) of 25 males and females are given below.

S.No.	Males	Females	S.No.	Males	Females
1	59	60	14	59	56
2	59	60	15	58	59
3	55	56	16	60	57
4	55	55	17	62	58
5	61	56	18	62	56
6	58	60	19	55	57
7	60	59	20	58	59
8	57	54	21	55	55
9	57	55	22	60	57
10	55	55	23	60	56
11	58	53	24	54	58
12	60	55	25	56	55
13	58	54	—	—	—

Perform t-test and investigate whether there is any significant difference between the MSS of males and females.

8.3 The blood cholesterol of 20 patients has been observed before and after giving a special diet that could reduce the cholesterol level. The following are the values recorded.

Before diet	After diet
250	220
217	170
253	255
240	210
220	230
218	225
227	213
254	204
212	210
230	197

Using the Data Analysis Pak, carry out an appropriate test to check whether the average cholesterol level has decreased after taking the special diet.

8.4 The shoot length of a plant whose seeds were subjected to ultrasound treatment for 15, 30, 45 and 60 minutes has been recorded by a researcher as follows. The number of observation under each treatment is not the same. Carry out ANOVA and test whether the average shoot length can be treated as the same at all levels of the treatment?

15-MTS	30-MTS	45-MTS	60-MTS
15	16	15	14
15	19	17	16
17	19	20	21
19	20	21	15
24	23	25	24
17	16	17	16
17	20	21	19
20	21	23	22
24	23		27
28	29		32
15	16		
	18		

8.5 Consider the data of the plant growth problem given in Chapter 3, enter the data in Excel worksheet in such a way that a two-factor ANOVA with replication can be carried out with this data. Perform ANOVA and comment on the findings.

25-DAS	V-407	V-410	V-450
Control	15,16,15,14,15	17,16,17,16,17	15,16,14,15,14
15-MTS	19,17,16,17,19	20,21,19,20,21	18,17,16,17,18
30-MTS	20,21,19,20,21	23,22,24,23,24	20,21,19,21,22
60-MTS	15,24,23,25,24	27,28,29,30,32	26,27,28,24,23

8.6 The following is a two-way table showing the distribution of players according to their level of aggression and performance in badminton.

Aggression	Performance		
	High	Low	Medium
High	3	8	13
Low	10	9	11
Medium	7	11	12

Enter this data in the Excel worksheet. Taking the null hypothesis that aggression and performance are independent, calculate the expected frequencies and prepare another two-way table in the Excel worksheet. Now use the Chi-square paste function and examine whether the null hypothesis is true. What is the observed significance level?

Correlation and Regression Analysis

One of the main objectives of a research study is to examine the relationships among different variables. The strength of relationship between any two variables is measured in terms of an index called *correlation coefficient*. When there are several interrelated variables in a problem, the study of these coefficients helps in picking up such variables which are closely associated with each other. Excel provides a module to compute these correlation coefficients. The price and demand of a product, the body height and body weight of a person, the reading hours of a student and his performance in the examination are some examples of correlated variables.

Another related area in the study of relationships is called the *regression analysis*. The objective here is to develop a mathematical model that relates one variable with other or several other variables. It would then be possible to predict or estimate the value of one variable by knowing another variable (or a group of variables). As an example, given the price of a commodity we would be able to predict the quantity that can be sold at that price.

Excel has a separate module to carry out regression analysis. A detailed discussion on correlation and regression cannot be discussed within the scope of this book. Hence we discuss only the nature of the problem and study the Excel output for it.

9.1 CORRELATION ANALYSIS

The study of the relationship that exists among two or more random variables (like height and BMI of tribal men) is the main objective of the correlation analysis. It includes the identification and summary of such relationships. Pearson's correlation coefficient denoted by the letter r gives the intensity of the linear relationship between the two variables.

Let X and Y be two random variables representing two characteristics which are known to have some meaningful relationship. The first step in correlation analysis is the study of the *scatter diagram*, which plots the sample values of Y against the values of X. This can be created in Excel graphs by choosing the item Scatter diagram. If the scatter shows a fairly linear trend, we can compute the Pearson's correlation coefficient r. Otherwise, a non-linear relationship has to be studied. If μ_x and μ_y denote the means of X and Y respectively then the formula for the correlation coefficient is given by

$$r = \frac{n\Sigma XY - (\Sigma X)(\Sigma Y)}{\sqrt{[n\Sigma X^2 - (\Sigma X)^2][n\Sigma Y^2 - (\Sigma Y)^2]}}$$

The value of r lies between -1 and $+1$. A positive value of r indicates a positive relationship and a negative value indicates a negative relationship. If $r = +1$, we say that the relationship is perfect and positive. Similarly a perfect negative relationship is indicated by $r = -1$. When the correlation is positive, it means that one variable is in sympathy with the other. For instance if $r = +0.65$ between X and Y, then if X increases, Y also increases and the relationship is stronger than the case in which $r = +0.23$.

The square of the correlation coefficient is a measure of the percentage of variance in one variable explained by the other. We call it r^2 value and it is used to judge the adequacy of the linear model to study the relationship. It is also called the *coefficient of determination* and its value lies between 0 and 1. When $r = 0.65$ we get $r^2 = 0.4225$ or 42.25%. It means that Pearson's correlation coefficient explains 42.25% of the variation in $Y(X)$ in terms of $X(Y)$. The rest is unexplained by correlation coefficient. A higher value of r^2 indicates a stronger linear relationship than a lower value of r^2.

Once the value of r is calculated, it is not the end of the job. It should be remembered that it is based only on the sample values. One natural question is whether the population values could have a similar correlation or could it be taken as zero? Is the observed value of r an occurrence by chance in the process of random sampling or is there really some correlation between the variables in the population? Answers to these questions follow from a test of significance of r. Here is an example.

EXAMPLE 9.1 Data relating to the resistance in ohms (X) applied on an electric component and the failure time in minutes (Y) of 24 electronic components has been obtained by an engineer as a part of quality control study. The data is shown in Table 9.1.

Analysis We shall now calculate the correlation coefficient between X and Y after examining the scatter diagram.

First the data should be entered in the first two columns with the headings as X and Y (in the first row). Using Excel graphs, the scatter diagram appears as shown in Figure 9.1. The Excel Paste function, which is available in the main menu, can be used to compute

Table 9.1 Data on Resistance and the Failure Time of Components

Sample	X	Y	Sample	X	Y
1	43	32	13	37	30
2	29	20	14	36	36
3	44	45	15	39	33
4	33	35	16	36	21
5	33	22	17	47	44
6	47	46	18	28	26
7	34	28	19	40	45
8	31	26	20	42	39
9	48	37	21	33	25
10	34	33	22	46	36
11	46	47	23	28	25
12	45	36	24	48	45

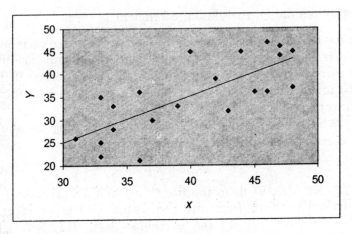

Fig. 9.1 Scatter diagram showing a positive correlation.

the correlation coefficient. We can either use the mouse to click on f_x or we can use the key sequence

Insert ▸ Function ▸ Statistical ▸ Correl

A dialogue box appears giving the range of X and Y values separately. If everything is correct, click on OK. The r-value obtained is 0.8085 and $n = 24$. Excel does not however give the p-value regarding the significance of the observed r. We have to calculate the p-value by carrying out the following steps:

1. Calculate

$$t_{cal} = r\sqrt{\frac{n-2}{1-r^2}}$$

where n is the number of data pairs and r is the observed correlation coefficient. This can be done in any cell of the Excel worksheet.

2. This gives $t_{cal} = 6.4401$
3. Select the Paste function and choose t-test among the statistical functions.
4. Type 6.4401 in the window against X.
5. Input the degrees of freedom as 22. (The degrees of freedom are $(n - 2)$.)
6. Then the t-distribution function returns the cumulative probability up to 6.4401 and gives the value as 1.76634E–06, which equals 0.000001766.
7. This is the required p-value, which can be approximated to 0.0000.

The null hypothesis is

H_0: The population correlation coefficient is zero.

If we set $\alpha = 0.01$, then the conclusion is that the observed correlation coefficient is highly significant at 1% level because p-value is almost equal to zero. In other words, the population correlation coefficient cannot be taken as zero and we say that the correlation coefficient is significant. ∎

Here is another case in which data from an existing file is used and the correlation coefficients between pairs of several variables are to be calculated.

EXAMPLE 9.2 Let us open the file C:\STATMAN\TRIBAL.DBF in Excel. We wish to find out the value of the correlation coefficient for each pair of the variables AGE, HT, WT, BMI, BMR and ENR.

Analysis Let us use the Data Analysis Pak and click on the item Correlation. As usual, the input is given by clicking on the window and blocking the columns having the variables in which we are interested.

It is necessary that all the desired columns must be adjacent to each. If we wish to include variables that are not found in the adjacent columns, we proceed as follows.

- Insert a blank column.
- Copy or drag the desired column to this position
- Repeat the above steps until all the required columns are brought to the adjacent positions.

With all this exercise we get the correlation coefficients as shown in Table 9.2. The correlation coefficient between a variable and itself is

Table 9.2 Correlation Matrix given by Excel.

	AGE	HT	WT	BMI	BMR	ENR
AGE	1					
HT	0.29802	1				
WT	0.19769	0.55816	1			
BMI	0.06017	0.02128	0.83943	1		
BMR	0.39803	0.66362	0.81839	0.55886	1	
ENR	0.31667	0.39772	0.71442	0.60693	0.83875	1

always 1 and hence all the values along the diagonal of the table are equal to 1. This table is called the *correlation matrix*. Excel does not show the significance of the observed correlation coefficient. However we can adopt the procedure given in Example 9.1 to find the significance level for each coefficient separately. ■

The correlation can also be understood by knowing what it does not mean. Correlation does not imply causal relationship. It only measures the strength of linear relationship between the variables. Sometimes the data may show a positive (or negative) relationship between the *shoe-size* and the *monthly income* of a person but it is non-sense. Such situations arise when the user attempts to work out the correlation coefficient without verifying whether the variables are related at all. The researcher has to determine whether the variables under consideration have any meaningful relationship at all.

9.2 SIMPLE REGRESSION ANALYSIS

Regression analysis is a statistical technique used to study the cause and effect relationship between one dependent variable (response) and one or more independent (predictor) variables. For instance, if Y denotes the quantity sold and X the selling price, then the relationship between Y and X can be established with the help of regression analysis. It provides a mathematical model relating to the response variable which is often called the *effect* with different explanatory variables called the *factors* or *causes*. The resulting model becomes a 'cause and effect' relationship and can be used to estimate the response at a chosen level of the factors.

A simple linear regression model is a mathematical equation of the type $Y = b_0 + b_1 X + \varepsilon$, where b_0 and b_1 are constants to be determined from the data and ε is called the *random error component*. This is also known as a *bivariate regression model*. Unless otherwise specified, we refer to linear regression alone when we use the word 'regression'.

If we have n pairs of data like (y_1, x_1), (y_2, x_2), ..., (y_n, x_n), we can estimate the values of b_0 and b_1 in such a way that the variance of the random error component is minimized. This is called ordinary least squares (OLS) method. The coefficient b_1 is called the regression coefficient given by the formula

$$b_1 = \frac{n\Sigma XY - (\Sigma X)(\Sigma Y)}{n\Sigma X^2 - (\Sigma X)^2}$$

The other constant b_0 is given by

$$b_0 = \overline{Y} - b_1 \overline{X}$$

where $\overline{Y}$ and $\overline{X}$ are the means of Y and X respectively.

The basic assumption in this regression model is that there is a known linear relationship between Y and X. If this is not true, we have to use a non-linear regression model. The scatter diagram helps judging whether a linear model is suitable or not. Thus a linear regression model is used only when the analyst is convinced that the relationship between the variables is like a straight line when plotted as a graph.

The adequacy of the linear model is judged with the help of a value called R^2-value. It is the square of the correlation coefficient and lies between 0 and 1. If the value of R^2 is high and close to '1' like 0.80 or 0.92, etc., we say that the linear model is a good fit.

An example of a linear regression model is $Y = 29.68 + 0.1344X$. Here $b_1 = 29.68$ and it is called the *intercept*. The regression coefficient is $b_0 = 0.1344$ and it is called the *slope* of the fitted line. The intercept gives the estimated value of Y when X is kept zero. It is the value of Y that holds good even if the effect of X is removed. We may also get a negative intercept, which means that Y will be positive only when X has some minimum positive value instead of zero.

If the regression coefficient has a positive sign, it means that X is positively correlated with Y. In other words, an increase in X by one unit leads to a marginal increase in Y by an amount of b_1 units. Similarly, a negative regression coefficient represents a negative relationship between Y and X. It may be noted that the sign of the correlation coefficient and that of the regression coefficient are one and the same.

We can use the regression model to predict the value of Y at a selected value of X. Substituting the value of X in the fitted model and simplifying the resulting expression gives a value of Y called the *predicted value* of Y or the *estimate* of Y. It is the average value of Y that can be obtained at the chosen value of X.

CORRELATION AND REGRESSION ANALYSIS

Besides this predicted value we have to provide the standard error (SE) of the estimate given by

$$s_{y,x} = \sqrt{\frac{\Sigma(y - \hat{y})^2}{n - 2}}$$

where n is the sample size and $\hat{y}$ is the estimate of Y corresponding to a value of X. Smaller the value of SE, more adequate is the fitted line in explaining the relationship.

Let us see how the Data Analysis Pak of Excel helps carrying out regression analysis of a given data.

EXAMPLE 9.3 The body weight and the BMI of 12 school going children are given in the following table.

Weight	BMI	Weight	BMI
15.0	13.35	32.0	15.65
26.0	16.12	18.0	13.85
27.0	16.74	22.0	16.07
25.0	16.00	20.0	12.8
25.5	13.59	26.0	13.65
27.0	15.73	24.0	14.42

Let us fit a simple regression model of BMI on weight and examine the results given by Excel.

Analysis We first note that the scatter diagram shown in Figure 9.2 indicates a positive relationship between BMI and the weight.

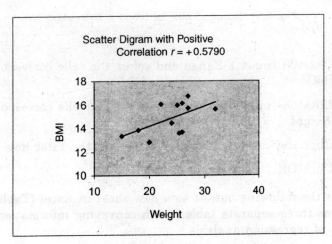

Fig. 9.2 A case of positive correlation.

In order to fit the regression model, we proceed as follows.

- Open Data Analysis Pak
- Select the item Regression and click on OK
- A dialogue box appears as shown in Figure 9.3

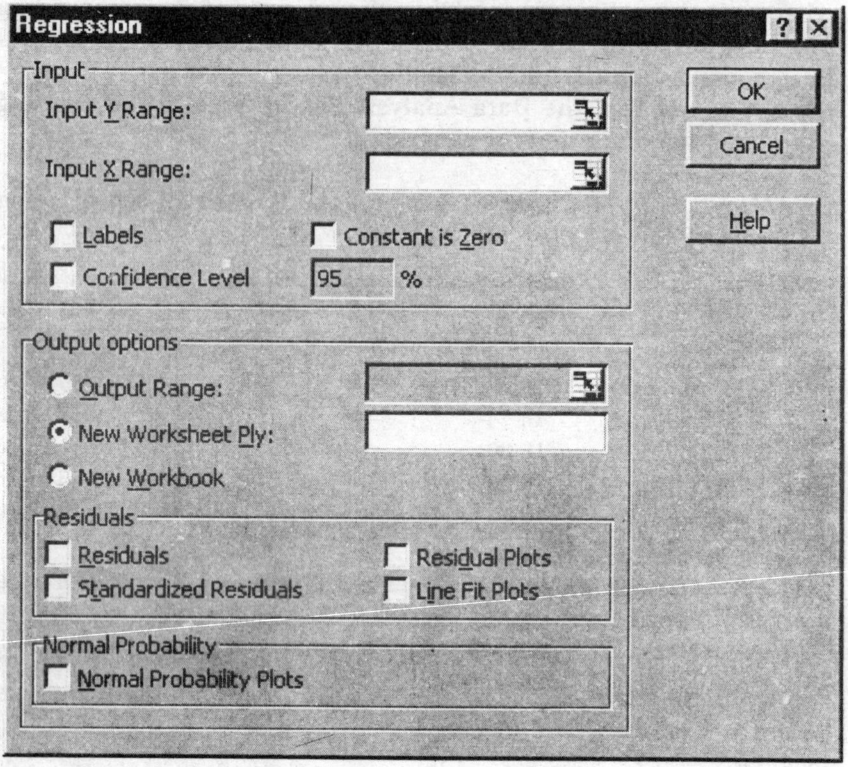

Fig. 9.3 Dialogue box for fitting a regression line.

- Click on Input *Y*-Range and select the cells corresponding to BMI
- Click on Input *X*-Range and select the cells corresponding to Weight
- Click the check box against Labels in the First Row
- Click OK.

We get the following output as a new sheet in Excel (Table 9.3). It contains three separate tables, each conveying information on one aspect of regression analysis.

Table 9.3 Excel Output for Regression Analysis

SUMMARY OUTPUT

Regression	Statistics
Multiple R	0.57902
R square	0.33527
Adjusted R square	0.26880
Standard error	1.15507
Observations	12

ANOVA

	DF	SS	MS	F	Significance F
Regression	1	6.7291	6.7291	5.0437	0.0485
Residual	10	13.3418	1.3342		
Total	11	20.0709			

	Coefficients	Standard Error	t Stat	p-value	Lower 95%	Upper 95%
Intercept	10.73487	1.85405	5.78995	0.00018	6.60378	14.86596
Weight	0.17096	0.07612	2.24581	0.04852	0.00135	0.34058

The first part gives the multiple R as 0.57902, which is the correlation coefficient between the body 'weight' and BMI. Normally in case of bivariate relations, the correlation coefficient is denoted by 'r', and 'R' is used in case of more than two variables. Excel output however uses R for both.

The value of R^2 is 0.3352, which means that about 33.52% variation in BMI can be explained by 'weight' through this linear model. This is apparently low because more than 67% of variation remains unexplained. There could be several reasons for this and one of them is that there might be some other influencing variables that have not been included in the present model.

Since the R-value is only an estimate, Excel gives its standard error and the number of observations used for the analysis.

Along with the R^2 value, Excel provides another analysis called ANOVA. It is a test of significance for comparing the variance due to regression with that of the unexplained variation (called *residual*). The F-value shown in this table gives the statistics for the variance ratio test of the regression model. The significance of F, which is given as 0.0485, is the p-value of the F-test carried out in ANOVA. If this value is less than 0.05 we say that the regression is statistically significant at 5% level of significance. Here, the regression is significant which means that the relationship is not an occurrence by chance.

The third and last part of the Excel output is related to the main results of regression. By reading the contents in the first and second columns of this table we find that b_0 is the intercept with a value of 10.7349 and b_1 is the regression coefficient due to weight with a value of 0.1710. The regression coefficient is positive, which shows that the BMI is positively related to weight. Excel conducts t-tests for these coefficients because they are only estimates and the p-value corresponding to them shows that both the coefficients are statistically significant.

This regression output can be written as a mathematical equation

$$\text{BMI} = 10.7349 + 0.1710 * \text{Weight}$$

Suppose the body weight of one student is known as 25 kg. Using the above equation, the estimated BMI is 15.01. Since this is only an estimate we have to interpret it as the average BMI corresponding to the given weight assuming that other parameters are unchanged. This estimate has a standard error of 1.5507.

We also observe that the sign of the correlation coefficient and that of the regression coefficient are the same, as they should be. Further the p-value corresponding to the F-test in the ANOVA and the t-test in the table of coefficients corresponding to weight are one and the same. ■

In the following example, we consider a case of a negative correlation.

EXAMPLE 9.4 The following data refers to the daily sale of tomatoes (in kg) at different prices (in rupees) observed on different days in a market.

Price	Quantity sold	Price	Quantity sold
4.5	125	5.5	130
5.5	115	6.5	120
4.5	140	5.0	130
4.5	140	5.5	100
4.0	150	6.0	105
5.5	150	4.5	150

Let us carry out linear regression analysis for this data.

Analysis The linear regression procedure of Excel gives the following results.

$R = 0.6345$
$R^2 = 0.4025$
F-value = 6.7384
p-value of $F = 0.0267$
Model: Quantity sold = 205.1375 − 14.7423 * Price

The fitted model has a statistically significant R^2 value. The coefficient of price is negative which means that one unit increase in price would decrease the average sales by about 14.74 units.

Here is another situation in which the variables are uncorrelated.

EXAMPLE 9.5 The body weight and the amount of calcium intake have been measured for 20 individuals in connection with a nutrition study on school going children. The data is shown in Table 9.4.

Table 9.4 Data on Calcium Intake and Body Weight of 20 Students

S.No.	Calcium	Weight	S.No.	Calcium	Weight
1	372.3	15.0	11	342.3	26
2	253.0	26.0	12	378.0	24
3	309.0	27.0	13	329.6	26
4	383.3	25.0	14	359.0	29
5	342.3	25.5	15	338.6	30
6	277.3	27.0	16	352.0	26
7	353.0	32.0	17	290.3	30
8	324.6	18.0	18	348.3	23
9	247.3	22.0	19	379.3	30
10	333.3	20.0	20	348.6	28

Analysis The scatter diagram shown in Figure 9.4 shows no trends either increasing or decreasing and the correlation coefficient is found to be –0.0087. This coefficient is almost zero.

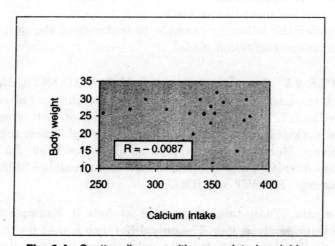

Fig. 9.4 Scatter diagram with uncorelated variables.

If we fit a regression line of body weight on calcium intake, we get the model

$$\text{Weight} = 25.788 - 0.00094 * \text{Calcium}$$

As expected, this equation has a regression coefficient very close to zero. In the following section, we study the relationship involving more than two variables.

9.3 MULTIPLE LINEAR REGRESSION

Multiple regression is used when the dependent variable, which is often called the 'response', is influenced by more than one explanatory variable. The general form of a multiple linear regression model with k explanatory variables is given by

$$Y = b_0 + b_1 X_1 + b_2 X_2 + \cdots + b_k X_k + \varepsilon$$

These k regression coefficients are estimated by the method of least squares and Excel automatically takes care of this estimation. The regression coefficient b_i corresponding to X_i gives the marginal partial contribution of the variable X_i to the response Y while the other variables are held at constant level. As done in the case of bivariate regression, here also the Excel Data Analysis Park asks for the input range for Y and X values. In the provision made for selecting the X-range, we have to select all the adjacent columns, which contain the explanatory variables. If the data requires different variables which are not available in adjacent columns, we have to bring them together by using the commands Insert ▸ Column and then dragging the required columns to one place.

Consider the following example to understand the output of a multiple linear regression model.

EXAMPLE 9.6 Consider the data file C:\STATMAN\TRIBAL.DBF and copy the first 30 rows of the data corresponding to the columns BMR, AGE, HT, WT and BMI to a new worksheet. Let us save this in a new workbook named as MLR (to mean multiple linear regression data). These 30 observations appear as shown in Table 9.5.

Let us develop a multiple linear regression model of BMR on the variables age, HT, WT and BMI.

Analysis To do this, we proceed as done in Example 9.5 and use the Data Analysis Pak. The only difference is that for the range against the X-value, we block all the cells corresponding to all columns

CORRELATION AND REGRESSION ANALYSIS

Table 9.5 Portion of Data Extracted from TRIBAL.DBF

BMR	AGE	HT	WT	BMI
1459.3	21	158	51.0	20.43
1474.6	21	152	52.0	22.51
1413.4	22	159	48.0	18.99
1451.6	23	157	50.5	20.49
1551.1	24	157	57.0	23.12
1597.0	25	153	60.0	25.63
1352.2	25	156	44.0	18.08
1466.9	25	158	51.5	20.63
1581.1	26	159	57.0	22.55
1535.8	27	158	56.0	22.43
1505.2	27	154	54.0	22.77
1566.4	28	158	58.0	23.23
1581.7	29	161	59.0	22.76
1558.8	29	159	57.5	22.74
1453.2	30	157	49.5	20.08
1470.6	30	156	51.0	20.96
1505.4	32	155	54.0	22.48
1528.6	35	158	56.0	22.43
1569.2	37	154	59.5	25.09
1482.2	38	157	52.0	21.10
1401.0	40	159	45.0	17.80
1493.8	41	160	53.0	20.70
1447.4	44	156	49.0	20.13
1459.0	46	155	50.0	20.81
1470.6	49	150	51.0	22.67
1098.7	18	158	41.0	16.42
1201.6	19	159	48.0	18.99
1157.5	19	152	45.0	19.48
1054.6	20	146	38.0	17.83
1157.5	21	155	45.0	18.73

under age, HT, WT and BMI. The output of the regression in simplified form is presented below.

Regression statistics	
Multiple R	0.9328
R square	0.8701
Adjusted R square	0.8493
Standard error	58.0908
Observations	30

ANOVA

Source	DF	SS	MS	F	Significance F
Regression	4	565134.0781	141283.5195	41.8674	0.0000
Residual	25	84363.6685	3374.5467		
Total	29	649497.7467			

	Coefficients	Standard error	t Stat	p-value
Intercept	−2500.4921	4217.5492	−0.5929	0.5586
Age	4.0209	1.3163	3.0546	0.0053
HT	17.2929	27.1775	0.6363	0.5304
WT	1.0187	42.8418	0.0238	0.9812
BMI	50.5535	103.5823	0.4881	0.6298

In this output, we have not shown the 95% confidence intervals for the regression coefficients. They are required only when the researcher wants to carry out further analysis with the regression coefficients and draw inferences from them.

Now the regression model can be stated as

$$BMR = -2500.4921 + 4.0209 \text{ (Age)} + 17.2929 \text{ (HT)}$$
$$+ 1.0187 \text{ (WT)} + 50.5535 \text{ (BMI)}$$

R^2 is 0.8701, which is about 87% of BMR can be explained in terms of age, HT, WT and BMI of a person through this linear model. We also see that all the explanatory variables have positive relationship with BMR. These regression coefficients are however not statistically significant except that of age, though the F-test in ANOVA shows that the overall regression is significant at 0.01 level (p-value is almost zero). The meaning of the regression coefficient can be understood as follows: If the age increases by one year, the BMR can be estimated to increase by 4.0209 units at fixed values of the other factors like HT, WT and BMI. ∎

Consider another example related to a problem in marketing research.

EXAMPLE 9.7 The sale of a product in lakhs of rupees (Y) is expected to be influenced by two variables namely the advertising expenditure X_1 (in '000 Rs.) and the number of sales persons (X_2) in a region. Sample data on 8 regions of a state has given the following results.

Area	Y	X_1	X_2
1	110	30	11
2	80	40	10
3	70	20	7
4	120	50	15
5	150	60	19
6	90	40	12
7	70	20	8
8	120	60	14

We wish to establish a relationship between the sales and the other two variables.

Analysis As done in the previous exercise, the Data Analysis Pak options give the following output.

SUMMARY OUTPUT

Regression statistics	
Multiple R	0.9587
R square	0.9191
Adjusted R square	0.8867
Standard error	9.5935
Observations	8

ANOVA

	DF	SS	MS	F	Significance F
Regression	2	5227.3256	2613.6628	28.3986	0.0019
Residual	5	460.1744	92.0349		
Total	7	5687.5			

	Coefficients	Standard error	t Stat	p-value
Intercept	16.8314	11.8290	1.4229	0.2140
X_1	−0.2442	0.5375	−0.4543	0.6687
X_2	7.8488	2.1945	3.5766	0.0159

Now the regression model is

$$Y = 16.8314 - 0.2442 * X_1 + 7.8488 * X_2$$

Since the R^2 is 0.9587 and the ANOVA shows that the F-ratio is significant, this model can be taken as a good-fit in explaining the sales in terms of the other two variables. Now, let us try to know the relative importance of X_1 and X_2. The regression coefficients show that X_2 has a positive contribution to Y while X_1 has a negative contribution. It however does not mean that an increase in X_1 (the advertisement expenditure) leads to a decrease in Y (sales). It only means that, as per the data available, any further increase in advertisement expenditure leads to a fall in sales if we keep the number of sales persons fixed. A different way of understanding the relative importance of predictor variables is to calculate the *beta coefficients*, which are the regression coefficients obtained when all the variables are converted into standardized form, so that each variable will have a mean zero and standard deviation 1. However, Excel does not directly provide these coefficients.

Thus multiple linear regression can be used to establish relationships among the variables of interest.[1]

9.4 DIAGNOSTIC ANALYSIS OF REGRESSION

The R^2 value is the commonly used measure of the adequacy of the linear model. We can test for its statistical significance with the help of ANOVA and understand the p-value. Now a different question arises for the researcher. If the regression model is used to predict the values of Y at the known values of X, will these estimates match with the actual Y values? Will there be any difference? If so what would be the average error? Excel resolves these issues as a part of the regression analysis.

Consider the regression given in the Example 9.3. While selecting the output options from the dialogue box, let us select all the items, Residuals, Line Fit and Normal Probability Plots. Then we get the following details.

Residuals and residual plots

The difference between the actual and the predicted value of Y from a given model is called the *residual*. These residual are standardized so that their mean is zero and standard deviation is 1. If the model is a good fit to the data, we can expect these residuals to be close to zero. However they do differ and Excel shows these differences below. It is called the *residual output*.

Residual Output

Observation	Predicted BMI	Residuals	Standard residuals
1	13.2993	0.0507	0.046036
2	15.17988	0.940119	0.853636
3	15.35084	1.389158	1.261366
4	15.00892	0.991081	0.89991
5	15.0944	−1.5044	−1.36601
6	15.35084	0.379158	0.344278
7	16.20565	−0.55565	−0.50454
8	13.81219	0.037815	0.034336
9	14.49603	1.573967	1.429175
10	14.15411	−1.35411	−1.22954
11	15.17988	−1.52988	−1.38914
12	14.83796	−0.41796	−0.37951

Excel also provides the plot of these standardized residuals as shown in Figure 9.5.

---

[1]For more details on the interpretation of regression results, the reader is referred to Kachigan (1986) and Johnson (1995).

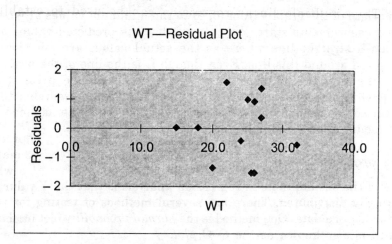

Fig. 9.5 Plot of standardized residuals.

Line fit plot

The plot of the actual and predicted values obtained from the regression model is shown in line fit graph. This is very useful to judge the adequacy of the model. If many of the predicted values differ from the actual values we may consider the model as not a good representative of the relationship. This plot is shown in Figure 9.6.

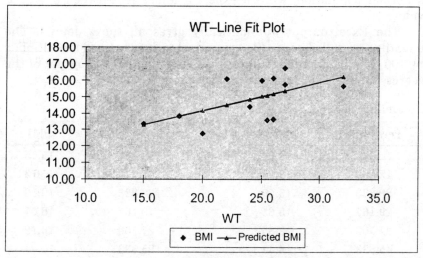

Fig. 9.6 The line fit plot of BMI against WT.

This graph shows the plot of BMI against the values of the weight. Both the actual and the plotted values are shown on the same chart. We have readjusted the scale of both X- and Y-axes in order to shift the origin of the graphs from (0, 0) to the minimum values at which the measurements start. We observe that the predicted values are all on a straight line where as the actual values are not closely scattered around this line. Even though it is the line of the best fit, still the available data could not give more accurate predictions than what is observed here. It is an indication to the researcher that either some more data or a few other variables could be included in the model in order to make a good line-fit.

The probability plots

One of the assumptions in regression analysis is that the Y value is normally distributed. There are several methods of testing for the normality of a data. One method is the *normal probability* plot method, which is also known as the *Q–Q plot*.

It is a plot of the observed data against the normal quartiles. The method is as follows:

- Let there be n data values
- Arrange the data in the increasing order and denote them by $X_1, X_2, ..., X_n$.
- Calculate $q_1 = (1 - 0.5)/n$, $q_2 = (2 - 0.5)/n$, ..., $q_n = (n - 0.5)/n$
- Plot the pairs of data $(q_1, x_1), (q_2, x_2), ..., (q_n, x_n)$.

This gives a normal probability plot. If the data has truly come from a normal population, we can expect a linear fit between q_i and x_i values.

The Excel output automatically gives all these details. The quantiles are called *percentiles*, since each value obtained is multiplied by 100 for a better understanding. The following is the output of the regression regarding the probability plots.

Probability Output

Percentile	BMI	Percentile	BMI
4.167	12.80	54.167	15.65
12.500	13.35	62.500	15.73
20.833	13.59	70.833	16.00
29.167	13.65	79.167	16.07
37.500	13.85	87.500	16.12
45.833	14.42	95.833	16.74

The plot of the BMI values against different percentiles is shown in Figure 9.7.

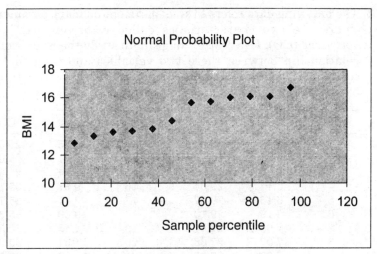

Fig. 9.7 Normal probability plot of BMI values.

We find from this graph that the variable BMI is approximately normally distributed because the plotted values have not deviated much from the straight line. This method works well when the number of values is moderately large.[1]

We end this chapter with the observation that Excel features can be efficiently used in explaining the relationships among variables.

REFERENCES

1. Kachigan Sam Kash (1986): *Statistical Analysis—An Interdisciplinary Introduction to Univariate and Multivariate Methods*, Radius Press, New York.
2. Richard A. Johnson and Dean W. Wichern (1992): *Applied Multivariate Statistical Analysis*, 3rd ed., Ch. 4, Prentice-Hall of India, New Delhi.
3. Richard A. Johnson (1995): *Miller and Freund's Probability and Statistics for Engineers*, Prentice-Hall of India, New Delhi.

DO IT YOURSELF

9.1 Construct a scatter diagram between the variables EIN and EEX of the data corresponding to women given in TRIBAL.DBF. Is the correlation positive or negative?

---

[1]More information on the Q–Q plot can be seen in Johnson and Wichern (1992).

9.2 The following data refers to an experiment on plant parameters. There are two characteristics called *embryonix* (EA) and *cotlydons* (CO). Use scatter diagram to study the nature of the relationship between these two variables and then find the correlation coefficient. What do you notice regarding the strength of the linear relationship between the variables?

S.No.	EA	CO	S.No.	EA	CO
1	9.12	20.90	13	11.75	25.54
2	9.18	20.86	14	11.61	25.62
3	9.20	20.91	15	11.54	25.59
4	6.14	29.04	16	9.81	35.64
5	6.15	29.07	17	9.71	35.74
6	6.19	29.10	18	9.69	35.68
7	7.05	27.71	19	14.48	30.54
8	7.07	27.68	20	14.54	30.65
9	7.03	27.59	21	14.60	30.58
10	10.81	16.08	22	16.81	20.08
11	10.74	16.05	23	16.74	20.10
12	10.85	16.10	24	16.78	20.15

9.3 The following data gives the number of hours (Y) required for drying of a paint with different doses of varnish added (X) in grams. Observe that the scatter shows a clear non-linear relationship between X and Y.

X	Y	X	Y
0	12.0	6	6.4
1	10.5	7	6.8
2	9.4	8	7.1
3	8.0	9	7.4
4	7.2	10	8.0
5	7.0	—	—

Compute the correlation coefficient and comment.

9.4 The following table gives data on three variables Y, $X1$ and $X2$, where $X2$ takes only two values 0 and 1. It is known as a *dummy* variable and used to account for the effect of categorical variable like gender, yes/no type answers, etc.

Y	X1	X2	Y	X1	X2
97	7	0	33	2	0
57	4	1	118	8	0
78	5	0	65	5	1
10	1	0	25	2	1
75	5	0	71	5	0
62	4	1	105	7	1
101	7	1	17	1	0
27	2	0	49	4	0
53	4	0	68	5	0

Show that Excel gives the linear regression model as

$$Y = -1.9719 + 14.6870 * X1 + 0.1514 * X2$$

What is the R^2 value? Is it significant?

9.5 One of the problems in regression analysis is about *multicollinearity*, which arises when the explanatory variables are correlated with each other. Such variables do not show their independent influence on the response variables. The presence of multicollinearity can be identified by examining the correlation matrix. Excel has a module in the Data Analysis Pak to compute the correlation matrix. If any of the explanatory variables has significant correlation coefficients, it is an indication of multicollinearity. Create a multiple regression problem and compute the correlation matrix.

FoxPro Programs for Quick Statistics

In this chapter, a few FoxPro programs are given for computing basic statistics, so that the user can use the PC for data handling. Whenever standard software is available, we however recommend its use instead of spending time on writing programs. However, it proves very useful if one can generate a custom-designed output for the problem under consideration. In such cases, these FoxPro programs provide an insight into the art of writing simple programs.

Each program should be typed in FoxPro after giving the command MODI COMM {file name}. It gives a screen for typing the commands. After typing all the commands, we have to save the program by using the key combination Ctrl + W. This creates the program with file name given by us, and the extension is PRG. When the DIR command is used, we do not find these PRG files in the list. To see the available programs, we have to type the command DIR *.PRG.

It is suggestive to keep all the programs in a separate directory say MYPROGRAMS. We can open FoxPro from this directory and enter all the programs.

In order to run any program, the command is DO {program name}. For instance, if we have a program BLANK.PRG, we can run it by typing the command DO BLANK. If we wish to make any changes in the program we have to use MODI COMM {program name}. For example, MODI COMM BLANK.

In the following sections we discuss some simple programs.

10.1 PROGRAM 1: CREATING A BLANK DATABASE FILE

Suppose we wish to create a file named MYPHONE and wish to enter the names and telephone numbers of persons. We can create a file with the fields SNO, NAME and PHONE. Assume that we have 100 records to enter in this file. One way of entering the data is to type all the fields of each record at a time. Alternatively, we can create a blank file with 100 records and replace the serial number (SNO) with the record number. Then it would be convenient to edit each field and enter data in that field alone. We can also create a new field and enter data in that field alone for all the 100 records. This method however does not prove useful if we have an unknown number of records. Following is a program for this purpose:

```
*PROGRAM NAME: BLANK
SET TALK OFF
CLEA ALL
CLEA
ACCEPT 'FILE NAME' TO FF
USE &FF
INPUT 'NUMBER OF BLANK RECORDS YOU NEED' TO M
I = 1
DO WHILE I <= M
APPE BLANK
REPL ALL SNO WITH RECNO()
I = I+1
ENDDO
SET TALK ON
CLOSE ALL
*END OF THE PROGRAM
```

When we run this program with DO BLANK command, it asks for the name of the file and the number of blank records to be given. We have to type the name of the file along with its path. For instance, if our file is E:\HOSTEL\STUD we have to type the file name in full like this and not simply STUD. This program can help creating files for structured data entry like analysis of data from designed experiments.

Another program for computing the basic statistics for the data given in any FoxPro file is discussed below:

10.2 PROGRAM 2: DESCRIPTIVE STATISTICS (DSTAT)

This program can be used to compute the basic statistics mean, median, mode, standard deviation, skewness and kurtosis of the data in a numeric field of a FoxPro file. After typing the file name, all the variables (fields) available in that file are displayed. The program asks for the name of the variable for which the basic statistics should be computed. Once we type the name correctly, we get the results on the screen. A hard copy of the results can also be obtained on the printer. The detailed program DSTAT is given below.

```
*PROGRAM NAME: DSTAT
SET TALK OFF
SET DATE BRITISH
SET STAT ON
SET FIXED ON
```

```
SET DECIMALS TO 4
CLEA ALL
CLEA
DELE FILE TT.DBF
@8,25 TO 14,55
@9,28 SAY ' WELCOME TO DSTAT'
@11,28 SAY ' DO IT YOURSELF '
@13,28 SAY 'INTERACTIVE FOXPRO PROGRAM '
?
?
WAIT
CLEA
ACCEPT 'FILE NAME TO USE ? ' TO FF
USE &FF
USE TRIBAL
CLEA
?' FILE NAME : ', FF
?
FC = FCOUNT()
SS = 'Y'
R = 1
DO WHILE SS = 'Y' .OR. SS = 'Y'.AND. R <= FC
?
?'LIST OF VARIABLES: '
?
FC = FCOUNT()
I = 1
S = 0
FOR I = 1 TO FC
FN = FIEL(I)
FN1 = 10-LEN(FN)
??FN,SPAC(FN1)
S = S+1
IF S < 7
??
ELSE
?
S =0
ENDIF
ENDFOR
?
ACCEPT ' WHICH VARIABLE TO SELECT ? ' TO V
V = UPPER(V)
?
SORT ON &V TO TT
USE TT
CALC MIN(&V) TO N1
CALC MAX(&V) TO N2
LINE = REPL('-',60)
GO BOTT
NN= RECNO()
CLEA
```

```
?
GO TOP
Y1 = MOD(NN,4)
Y2 = MOD(NN,2)
Y3 = MOD(3*NN,4)
IF Y1 = 0
LOCA FOR RECNO() = NN/4
J1= RECNO()
X11 = &V
SKIP
X12 = &V
Q1 = (X11+X12)/2
ELSE
L1 = INT((NN+1)/4)
GO L1
Q1= &V
ENDIF
GO TOP
IF Y2 = 0
L2 = INT(NN/2)
GO L2
X21 = &V
SKIP
X22 = &V
MEDIAN = (X21+X22)/2
ELSE
L2 = INT((NN+1)/2)
GO L2
MEDIAN = &V
ENDIF
GO TOP
IF Y3 = 0
L3 = INT(3*NN/4)
GO L3
X13 = &V
X23 = &V
Q3 = (X13+X23)/2
ELSE
L3 = INT((3*NN+1)/4)
GO L3
Q3 = &V
ENDIF
GO TOP
CALC AVG(&V) TO MEAN
CALC SUM(&V*&V) TO V2
CALC SUM(&V) TO V1
VAR = (NN*V2-V1*V1)/(NN*(NN-1))
SD = SQRT(VAR)
GO TOP
CALC SUM(((&V-MEAN)/SD)*((&V-MEAN)/SD)*((&V-MEAN)/SD)) TO SK1
R1 = NN/((NN-1)*(NN-2))
SK = SK1*R1
```

```
CALC SUM((&V-MEAN)*(&V-MEAN)*(&V-MEAN)*(&V-MEAN)) TO KU1
KU1 = KU1/(SD*SD*SD*SD)
KR1 = (NN*(NN+1))/((NN-1)*(NN-2)*(NN-3))
KR2 = (3*(NN-1)*(NN-1))/((NN-2)*(NN-3))
KURT = KR1*KU1-KR2
MO = (3*MEDIAN) - (2*MEAN)
SET PRINT ON
? SPAC(5), 'FILE NAME :', FF
??SPAC(10), 'VARIABLE :', V
?SPAC(5),'BASIC STATISTICS COMPUTED ON ', DATE(), 'AT',TIME()
?
?SPAC(5), LINE
?
?SPAC(5),'NUMBER OF CASES   '+' =',TRANS(NN,"######")
?SPAC(5),'MINIMUM           '+' =',TRANS(N1,"#####.####")
?SPAC(5),'MAXIMUM           '+' =',TRANS(N2,"#####.####")
?SPAC(5),'MEAN              '+' =',TRANS(MEAN,"#####.####")
?SPAC(5),'VARIANCE          '+' =',TRANS(VAR,"#####.####")
?SPAC(5),'STD.DEV           '+' =',TRANS(SD,"#####.####")
?SPAC(5),'MEDIAN            '+' =',TRANS(MEDIAN,"#####.####")
?SPAC(5),'MODE              '+' =',TRANS(MO,"#####.####")
?SPAC(5),'FIRST QUARTILE    '+' =',TRANS(Q1,"#####.####")
?SPAC(5),'THIRD QUARTILE    '+' =',TRANS(Q3,"#####.####")
?SPAC(5),'SKEWNESS          '+' =',TRANS(SK,"#####.####")
?SPAC(5),'KURTOSIS          '+' =',TRANS(KURT,"#####.####")
?
?SPAC(5), LINE
?
SET PRINT OFF
WAIT
CLEA
ACCEPT 'DO YOU WANT TO REPEAT WITH OTHER VARIABLES? (Y/N)' TO S
*ENDDO
SET PRINT ON
CLEA
IF SS = 'N'.OR. SS = 'N'
CLEA
@6,25 TO 8,55
@7,27 SAY 'THANK YOU FOR USING DSTAT'
CLOSE ALL
ENDIF
IF SS = 'Y'.OR. SS = 'Y'
CLEA
R = R+1
USE
DELE FILE TT.DBF
USE &FF
ENDIF
ENDDO
SET TALK ON
CLOSE ALL
*END OF THE PROGRAM
```

We can run this program by typing the command DO DSTAT or by selecting the RUN option in the main menu that appears on the top of the screen. The interactive program works as shown in Figure 10.1.

```
        WELCOME TO DSTAT
        DO IT YOURSELF
   INTERACTIVE FOXPRO PROGRAM

Press any key to continue...
```

Fig. 10.1 Opening message for DSTAT program.

Now FoxPro asks for the name of the file to be used. We then type TRIBAL. Suppose we wish to use a file, which is in some other directory. Then we have to type the file name along with the name of the directory. Then the program displays the list of variables on the screen as shown in Figure 10.2 for selecting the variable. Let us type the name of the variable as BMI.

```
FILE NAME: tribal

LIST OF VARIABLES:
  SNO    SEX    AGE    HT     WT     BMI    BMR
  PRO    FAT    CHO    CAL    IRON   VITA   VITC
  EIN    EEX    EBAL   CODE

Which Variable to select? _
```

Fig. 10.2 List of variables in the file as displayed by DSTAT.

The basic statistics for the variable BMI appears as shown in Figure 10.3. When we press any key the program asks the following question. We have to type 'Y' for Yes or 'N' for No. (Lower cases can also be used.)

```
Do you want to repeat with other variables?  (Y/N)
```

If we type 'Y' we get the message for selecting another variable. If we type 'N' the program ends a closing message appears on the screen.

```
FILE NAME: tribal              VARIABLE: BMI
BASIC STATISTICS COMPUTED ON  22/02/00 AT 19:40:43
─────────────────────────────────────────────────
NUMBER OF CASES = 150
MINIMUM = 14.6000
MAXIMUM = 26.3700
MEAN = 19.9299
VARIANCE = 4.993
STD.DEV = 2.1212
MEDIAN= 19.7500
MODE = 19.3903
FIRST QUARTILE = 18.5200
THIRD QUARTILE = 20.8100
SKEWNESS = 0.5793
KURTOSIS = 0.8103
─────────────────────────────────────────────────
Press any key to continue...
```

Fig. 10.3 Output from DSTAT.

10.3 PROGRAM 3: CREATING A FREQUENCY TABLE (FTAB)

The frequency distribution of the data values of a variable can be prepared with the help of a simple FoxPro program given below. The program is named as FTAB to mean frequency tabulation and it provides two ways of computing the frequencies. One is by choosing the class intervals and the number of classes automatically using the Sturge's formula and the other is to choose them according to the user's choice. The following is the listing of the program.

```
*PROGRAM TO COMPUTE THE FREQUENCY DISTRIBUTION
SET TALK OFF
SET STAT ON
CLEA ALL
CLEA
DELE FILE TT.DBF
ACCEPT 'FILE NAME TO USE ? ' TO FF
USE &FF
CLEA
DIME LB(20)
DIME UB(20)
DIME F(20)
?' FILE NAME : ', FF
?
SS = 'Y'
R = 1
DO WHILE SS = 'Y' .OR. SS = 'Y' .AND. R <= 30
```

```
?' LIST OF VARIABLES:
?
FC = FCOUNT()
I = 1
S = 0
FOR I = 1 TO FC
FN = FIEL(I)
FN1 = 10-LEN(FN)
??FN,SPAC(FN1)
S = S+1
IF S < 7
??
ELSE
?
S =0
ENDIF
ENDFOR
?
ACCEPT ' WHICH VARIABLE TO SELECT ? ' TO V
V = UPPER(V)
?
SORT ON &V TO TT
USE TT
CALC MIN(&V) TO N1
CALC MAX(&V) TO N2
LINE = SPACE(10)+REPL('-',42)
NN= RECCOUNT()
CLEA
?
?'LEAST VALUE OF THE DATA IS ',N1
?
?'HIGHEST VALUE OF THE DATA IS ', N2
?
ACCEPT 'DO YOU WANT THE CLASS INTERVAL AND LIMITS AUTOMATIC? (Y/N)'
TO YY
IF YY = 'N'.OR. YY = 'N'
?
INPUT 'LOWER LIMIT : ' TO LL
?
INPUT 'CLASS WIDTH : ' TO CI
I = CI
N1 = LL
K = INT((N2-N1)/CI)
ENDIF
IF YY = 'Y' .OR. YY = 'Y'
K1 = 1+3.322*LOG10(NN)
K = INT(K1)+1
I = (N2-N1)/K
ENDIF
CLEA
SET PRINT ON
IF YY = 'Y' .OR. YY = 'Y'
```

```
?' FREQUENCY DISTRIBUTION OF THE VARIABLE ',V+'.'
?' (INTERVALS ARE AUTOMATIC)'
ENDIF
IF YY = 'N' .OR. YY = 'N'
?' FREQUENCY DISTRIBUTION OF THE VARIABLE ',V+'.'
?' (INTERVALS ARE BY CHOICE OF THE USER)'
ENDIF
?LINE
?SPAC(15),' INTERVAL',SPAC(10),' FREQUENCY'
?LINE
GO TOP
FOR M = 1 TO K+1
LB(M) = N1+(M-1)*I
UB(M) = N1+M*I
F(M) = 0
DO WHILE .NOT. EOF()
IF &V >= LB(M) AND &V < UB(M)
F(M) = F(M)+1
ENDIF
SKIP
ENDDO
GO TOP
ENDFOR
TF = 0
FOR M = 1 TO K+1
TF = TF+F(M)
?SPAC(10),TRANS(LB(M),'#####.##'),'=',
TRANS(UB(M),'#####.##'),SPAC(10),TRANS(F(M),'###')
ENDFOR
?LINE
?SPAC(10),'TOTAL NUMBER OF CASES ', SPAC(0),TF
?LINE
?
SET PRINT OFF
ACCEPT ' DO YOU WANT TO TABULATE FOR ANOTHER VARIABLES ? (Y/N)'
TO SS
IF SS = 'N'.OR. SS = 'N'
CLEA
@6,25 TO 8,55
@7,27 SAY 'THANK YOU FOR USING FTABS'
ENDIF
IF SS = 'Y'.OR. SS = 'Y'
R = R+1
ENDIF
*SET PRINT ON
USE
DELE FILE TT.DBF
USE &FF
ENDDO
SET TALK ON
CLOSE ALL
*END OF THE PROGRAM
```

We can run this program using the command DO FTAB. The program displays the field names and then asks for the variable to select. Let us select the variable name as BMI and opt for the class intervals to be chosen automatically. Then the output appears on the screen as shown in Figure 10.4.

```
Frequency Distribution of the Variable BMI
         (Intervals are automatic)
```

Interval	Frequency
14.60–15.91	4
15.91–17.22	7
17.22–18.52	26
18.52–19.83	42
19.83–21.14	42
21.14–22.45	10
22.45–23.75	11
23.75–25.06	2
25.06–26.37	5
26.37–27.68	1
Total numbers of cases	150

Do you want to tabulate for another variables? (Y/N) –

Fig. 10.4 Frequency distribution with automatic class intervals.

```
Frequency Distribution of the Variable BMI
     (Intervals are by Choice of the User)
```

Interval	Frequency
14.00–16.00	4
16.00–18.00	19
18.00–20.00	57
20.00–22.00	47
22.00–24.00	16
24.00–26.00	6
26.00–28.00	1
Total numbers of cases	150

Do you want to tabulate for another variables? (Y/N) –

Fig. 10.5 Frequency table according to user-defined classes.

Suppose we wish to select the classes as per our choice. Then we type 'N' against the choice for the type of class intervals. The program asks for choosing the lower limit and the class interval. Let us select the lower limit as 14 and the class width as 2. Then we get the frequency table as shown in Figure 10.5.

10.4 PROGRAM 4: CROSS-TABULATIONS (CROSSTABS)

Another important problem of data analysis is that of cross-tabulations. We may wish to find out the number of cases falling under different categories of two attributes. This gives a bivariate frequency table. We have learnt (in Chapter 3) a method of counting the number of cases satisfying two conditions at a time and thereby preparing a small frequency table. Here is a FoxPro program to compute the cross-tabulations with respect to two categorical variables. The levels of these variables determine the number of rows and the columns in the table.

```
*PROGRAM TO COMPUTE THE CROSS-TABULATIONS
CLEA ALL
CLEA
SET TALK OFF
SS = 'Y'
I = 1
DO WHILE SS = 'Y' OR SS = 'Y' AND I <= 10
INPUT 'HOW MANY ROWS : ' TO R
INPUT 'HOW MANY COLUMNS : ' TO C
N = 0
CH = 0
DIME A(R,C)
DIME RT(R)
DIME CT(C)
DIME E(R,C)
CLEA
FOR I = 1 TO R
?'ENTER THE VALUES OF ROW - ',LTRIM(STR(I))
FOR J = 1 TO C
INPUT ' ' TO A(I,J)
ENDFOR
?
ENDFOR
FOR I = 1 TO R
RT(I) = 0
FOR J = 1 TO C
RT(I) = RT(I) + A(I,J)
ENDFOR
N = N+RT(I)
ENDFOR
```

```
FOR J = 1 TO C
CT(J) = 0
FOR I = 1 TO R
CT(J) = CT(J) + A(I,J)
ENDFOR
ENDFOR
FOR I = 1 TO R
FOR J = 1 TO C
E(I,J) = RT(I)*CT(J)/N
CH = CH + A(I,J)*A(I,J)/E(I,J)
CHI = CH - N
ENDFOR
ENDFOR
CLEA
?
? 'DATA AND RESULTS FOR THE CHI-SQUARE TEST'
?
FOR I = 1 TO R
FOR J = 1 TO C
??A(I,J)
ENDFOR
?
ENDFOR
DF = (R-1)*(C-1)
? 'CHI-SQUARE VALUE = ', TRANS(CHI,'####.##')
? 'DEGREES OF FREEDOM = ',TRANS(DF,'###')
?
ACCEPT 'DO YOU WANT TO REPEAT WITH ANOTHER DATA ? ' TO SS
?
IF SS = 'Y' OR SS = 'Y'
I = I+1
ENDIF
CLEA
IF SS = 'N' OR SS = 'N'
@10,10 SAY 'THANK YOU'
ENDIF
ENDDO
SET TALK ON
CLOSE ALL
*END OF THE PROGRAM
```

When this program is run we get prompt messages asking for the first and second categorical variables, one after the other.

In order to understand the utility of this program, let us consider the FoxPro file STRESS as discussed in Example 7.5. The sample data with about 10 records is shown in Figure 10.6. This file contains fields serial number (SNO), SEX, arts/science (AS), management type (MGT), birth order (BORD), mother's occupation (MOCC), father's occupation (FOCC) and FCOD to mean coded score of academic stress.

SNO	SEX	AS	MGT	BORD	MOCC	FOCC	FCOD
1	1	1	1	1	2	3	2
2	1	1	1	2	2	1	3
3	1	1	1	1	2	3	2
4	1	1	1	2	2	3	1
5	1	1	1	1	2	4	2
6	1	1	1	5	2	1	2
7	1	1	1	1	2	4	1
8	1	1	1	2	2	5	1
9	1	1	1	1	2	4	1
10	1	1	1	1	2	5	1
11	1	1	1	1	2	4	3
12	1	1	1	3	2	3	3
13	1	1	1	3	2	5	2
14	1	1	1	1	2	4	2

Fig. 10.6 Data file for cross-tabulations.

The students have been classified according to their level of academic stress into three groups namely high, moderate and low with respective numeric codes 1, 2 and 3. Similarly, the variables SEX, MOCC and FOCC also have codes indicating the level of occupation. As such, all the variables in this file are categorical ones. We now wish to classify all the 480 students according to SEX and FCOD.

To do this, we run the program CROSS. The program asks for the first and second variables for classification and we can give them in the order whichever we like. Let us give SEX as the first variable and then FCOD as the second variable. The programs provides:

(a) the number of cases falling in each group,
(b) the percentage of cases and
(c) the expected number of cases if FCOD is hypothetically independent of SEX.

This is one requirement when the user wants to know whether the two variables have any association between them. This is tested by chi-square test. Now the output appears shown in Figure 10.7.

In this way we can use the program CROSS to create cross-tabulations over categorical variables. If we look at the results, we note that for SEX = 1 and FCOD = 1 the expected number of cases is 58.57 or 59 while the observed cases is only 37. The chi-square test compares this difference and provides a decision rule to accept or reject whether the assumption of independence is supported or not supported by the available data. We have to compare this chi-square value with the table value for 1 degree of freedom at 5% or 1% level of significance.

```
CROSS-TABULATION OF sex X fcod
                ACTUAL        % CASES        EXPECTED
                COUNT                        COUNT
sex = 1
    fcod = 1     37            7.71%          58.50
    fcod = 2    140           29.17%         121.50
    fcod = 3     63           13.13%          60.00
    TOTAL      240           50.00%
sex = 2
    fcod = 1     80           16.67%          58.50
    fcod = 2    103           21.46%         121.50
    fcod = 3     57           11.88%          60.00
    TOTAL      240           50.00%

CHI-SQUARE = 21.7372        D.F. = 2

DO YOU WANT TO REPEAT WITH OTHER VARIABLES? (Y/N)
```

Fig. 10.7 Output of CROSSTAB program.

10.5 PROGRAM 5: TWO SAMPLE TESTS FOR MEANS

A simple FoxPro program to perform the t-test for comparing the means of two groups of data is given in this section. It can read the cases corresponding to the groups from the FoxPro file and performs the t-test. The only requirement is the user has to keep a data file for the critical values of the t-distribution, which have been prepared at 0.01 and 0.05 levels only. Let us keep this file in the directory C:\STATMAN. Since this is only an illustrative program, we have used this data file; otherwise we can compute the p-value itself. The following is the listing of the program TTEST.

```
*PROGRAM TTEST
set talk off
clea all
clea
LINE = REPL('-',50)
SELE A
ACCEPT 'FILE NAME : ' TO FF
use &FF
?'FILE NAME : ',FF
?
SET PRINT ON
?CHR(14),' STATISTICAL ANALYSIS OF '+FF
?
```

```
?LINE
SET PRINT OFF
fc = fcount()
ss = 'Y'
K = 4
do while ss = 'Y' .or. ss = 'y' and K <= fc
SET PRINT OFF
?'LIST OF VARIABLES: '
?
FC = FCOUNT()
I = 1
S = 0
FOR I = 1 TO FC
FN = FIEL(I)
FN1 = 10-LEN(FN)
??FN,SPAC(FN1)
S = S+1
IF S < 7
??
ELSE
?
S =0
ENDIF
ENDFOR
?
ACCEPT 'VARIABLE FOR TESTING: ' TO V
&V = UPPER(V)
ACCEPT 'GROUPING VARIABLE: ' TO G
&G = UPPER(G)
CLEA
?
CALC MIN(&G) TO G1
CALC CNT() TO N1 FOR &G = G1
?
CLEA
GO TOP
CALC AVG(&V) TO M1 FOR &G = G1
CALC VAR(&V) TO V1 FOR &G = G1
VV1 = V1*N1/(N1-1)
S1 = SQRT(VV1)
CALC MAX(&G) TO G2
CALC COUNT() TO N2 FOR &G = G2
CALC AVG(&V) TO M2 FOR &G = G2
CALC VAR(&V) TO V2 FOR &G = G2
VV2 = V2*N2/(N2-1)
S2 = SQRT(VV2)
D = ABS(M1-M2)
XX = (N1+N2-2)
S12 = (N1*VV1+N2*VV2)/XX
RR = (1/N1)+(1/N2)
SX = SQRT(S12)*SQRT(RR)
T = D/SX
```

```
SELE B
USE E:\STATMAN\TTABLE
IF XX <= 30
   LOCA FOR SNO = XX
   STORE AT5 TO TC5
   STORE AT1 TO TC1
 ELSE
   LOCA FOR SNO = 35
   STORE AT5 TO TC5
   STORE AT1 TO TC1
ENDIF
K = K+1
SELE A
?LINE
SET PRINT ON
?'T-TEST FOR THE MEANS OF ',+'"'+V+'"'+' ACCORDING TO
','"'+G+'"'
LL = 20-LEN(G)
?
?G +' =', LTRIM(STR(G1))
??SPAC(LL),LTRIM(G +' ='), LTRIM(STR(G2))
?
?'N1 = ', RTRIM(TRANS(N1,'#####'))
??SPAC(8),LTRIM('N2 = '), RTRIM(TRANS(N2,'#####'))
?'MEAN = ',TRANS(M1,'#####.##')
??SPAC(8),'MEAN = ',TRANS(M2,'#####.##')
?'S.D = ',TRANS(S1,'#####.##')
??SPAC(8),'S.D = ',TRANS(S2,'#####.##')
?
?'MEAN DIFFERENCE = ',LTRIM(TRANS(D,'###.##'))
?LTRIM('T- CAL = '),LTRIM(TRANS(T,'###.##'))
??SPAC(3),LTRIM('D.F = '), LTRIM(TRANS(XX,'###'))
??SPAC(3)
IF T <= TC5
??'T-CRI = ', LTRIM(TRANS(TC5,'###.##'))
?LTRIM('DIFFERENCE IS NOT SIGNIFICANT')
ENDIF
IF T > TC5
   IF T <= TC1
   ??'T-CRI = ', LTRIM(TRANS(TC5,'###.##'))
   ?LTRIM('DIFFERENCE IS SIGNIFICANT AT 5% LOS')
ELSE
   ??'T-CRI ', LTRIM(TRANS(TC1,'###.##'))
   ?LTRIM('DIFFERENCE IS SIGNIFICANT AT 1% LOS')
   ENDIF
ENDIF
?LINE
SET PRINT OFF
wait
SET PRINT ON
ENDDO
SET TALK ON
CLOSE ALL
```

Suppose we run this program with the file TRIBAL.DBF. We get the output as in Figure 10.8.

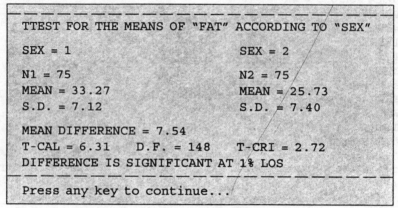

Fig. 10.8 Output of the program TTEST.

Since this program requires the data file TTABLE, we have given this data file in Table 10.1.

Table 10.1 Critical Values of the t-distribution

SNO	DF	AT5	AT1	SNO	DF	AT5	AT1	SNO	DF	AT5	AT1
1	1	12.706	0.000	15	15	2.131	2.947	29	29	2.045	2.756
2	2	4.303	9.925	16	16	2.120	2.921	30	30	2.042	2.750
3	3	3.182	5.841	17	17	2.110	2.898	35	35	2.030	2.724
4	4	2.776	4.604	18	18	2.101	2.878	40	40	2.021	2.704
5	5	2.571	4.032	19	19	2.093	2.861	45	45	2.014	2.690
6	6	2.447	3.707	20	20	2.086	2.845	50	50	2.008	2.678
7	7	2.365	3.499	21	21	2.080	2.831	55	55	2.004	2.669
8	8	2.306	3.355	22	22	2.074	2.819	60	60	2.000	2.660
9	9	2.262	3.250	23	23	2.069	2.807	70	70	1.994	2.648
10	10	2.228	3.169	24	24	2.064	2.797	80	80	1.989	2.638
11	11	2.201	3.106	25	25	2.060	2.787	90	90	1.986	2.631
12	12	2.179	3.055	26	26	2.056	2.779	100	100	1.982	2.625
13	13	2.160	3.012	27	27	2.052	2.771	120	120	1.960	2.617
14	14	2.145	2.977	28	28	5.248	2.763				

The user has to type this file into the directory E:\STATMAN in which the t-test program is maintained. The headers AT5 and AT1 represent the critical value at 5% and 1% level of significance.

10.6 PROGRAM 6: COMPUTATION OF CORRELATION COEFFICIENT

The program CORR helps calculating the correlation coefficient between the selected pair of variables from the file. It asks for the first and the second variables in the order and gives the r-value. The following is the program CORR.

```
*PROGRAM CORR
set talk off
clea all
clea
ACCEPT 'FILE NAME : ' TO FF
use &FF
DELE FILE TT.DBF
SELE A
CLEA
?'FILE NAME : ',FF
?
fc = fcount()
ss = 'Y'
K = 4
do while ss = 'Y' .or. ss = 'y' and K <= fc
?'LIST OF VARIABLES: '
?
FC = FCOUNT()
I = 1
S = 0
FOR I = 1 TO FC
FN = FIEL(I)
FN1 = 10-LEN(FN)
??FN,SPAC(FN1)
S = S+1
IF S < 7
??
ELSE
?
S =0
ENDIF
ENDFOR
?
N = RECCOUNT()
ACCEPT 'FIRST VARIABLE ? ' TO X
ACCEPT 'SECOND VARIABLE ? ' TO Y
CLEA
CALC AVG(&X) TO XBAR
CALC AVG(&Y) TO YBAR
CALC SUM((&X-XBAR)*(&Y-YBAR)) TO NUM
CALC SUM((&X-XBAR)*(&X-XBAR)) TO D1
CALC SUM((&Y-YBAR)*(&Y-YBAR)) TO D2
CALC CNT() TO N
DEN = SQRT(D1*D2)
CC = NUM/DEN
SET PRINT ON
?'MEAN OF '+X +'=', XBAR
?'MEAN OF ',+Y +'=', YBAR
?'CORR. COEFF ('+X +','+Y+')'+' = ',
```

```
LTRIM(TRANS(CC,'###.####'))
? 'DEGREES OF FREEDOM = ',LTRIM(STR(N-2))
DD = 1-(CC*CC)
T = CC*SQRT((N-2)/DD)
SELE B
USE TTABLE
LOCA FOR DF = (N-2)
IF T >= AT5
?? ' (SIG., AT 5%)'
ELSE
?? ' (NOT SIG AT 5% LOS)'
ENDIF
SELE A
SET PRINT OFF
?
ACCEPT 'DO YOU WANT TO REPEAT WITH OTHER VARIABLES ? (Y/N)' TO SS
if ss = 'N'.or. ss = 'n'
clea
@6,25 to 8,56
@7,28 say 'THANK YOU FOR USING CORR'
ENDIF
if ss = 'Y'.or. ss = 'y'
K = K+1
endif
enddo
SET TALK ON
CLOSE ALL
```

Let us use data file TRIBAL.DBF and select the variables FAT and CALCIUM for computing the correlation coefficient. The output appears as in **Figure 10.9**.

```
MEAN OF FAT = 29.50
MEAN OF CAL = 340.63
CORR COEFF (FAT; CAL) = 0.5700
DEGREES OF FREEDOM = 148 (SIG, AT 5%)

DO YOU WANT TO REPEAT WITH OTHER VARIABLES?   (Y/N)_
```

Fig. 10.9 Output of the program CORR.

10.7 PROGRAM 7: CHI-SQUARE TEST FOR INDEPENDENCE OF ATTRIBUTES

The following program can be used to enter the data of a contigency table and find the chi-square value.

```
*PROGRAM CHI
clea all
clea
set talk off
LINE = REPL('-',60)
```

```
CLEA
ss = 'Y'
I = 1
do while ss = 'Y' or ss = 'y'
ACCEPT ' FIRST VARIABLE (Along the rows) : ' TO FV
?
INPUT ' How many rows : ' to R
?
ACCEPT ' SECOND VARIABLE (Along the columns): ' TO SV
?
INPUT ' How many columns : ' to C
ENDIF
N = 0
CH = 0
DIME A(R,C)
DIME RT(R)
DIME CT(C)
DIME E(R,C)
CLEA
FOR I = 1 TO R
?'ENTER THE VALUES OF ROW - ',LTRIM(STR(I))
FOR J = 1 TO C
INPUT ' ' TO A(I,J)
ENDFOR
?
ENDFOR
FOR I = 1 TO R
RT(I) = 0
FOR J = 1 TO C
RT(I) = RT(I) + A(I,J)
ENDFOR
N = N+RT(I)
ENDFOR
FOR J = 1 TO C
CT(J) = 0
FOR I = 1 TO R
CT(J) = CT(J)+ A(I,J)
ENDFOR
ENDFOR
FOR I = 1 TO R
FOR J = 1 TO C
E(I,J) = RT(I)*CT(J)/N
CH = CH + A(I,J)*A(I,J)/E(I,J)
CHI = CH - N
ENDFOR
ENDFOR
CLEA
?
SET PRINT ON
?LINE
? 'DATA AND RESULTS FOR THE CHI-SQUARE TEST: ',FV,' X ',SV
```

```
?LINE
?
FOR I = 1 TO R
FOR J = 1 TO C
??A(I,J)
ENDFOR
?
ENDFOR
DF = (R-1)*(C-1)
? 'CHI-SQUARE VALUE = ', TRANS(CHI,'####.##')
?? SPAC(5),'D.F = ',TRANS(DF,'###')
?LINE
?
SET PRINT OFF
Accept 'Do you want to repeat with another data ? ' to ss
?
if ss = 'y' or ss = 'Y'
I = I+1
ENDIF
CLEA
if ss = 'N' or ss = 'n'
@10,10 SAY 'THANK YOU'
ENDIF
ENDDO
SET TALK ON
CLOSE ALL
```

Here is a sample output of a contingency table with 3 rows and 2 columns. The attribute shown along the rows is OPINION and the attribute shown along the columns is GENDER. Proceeding with the directions in the program we get the output as in Figure 10.10.

```
DATA AND RESULTS FOR THE CHI-SQUARE TEST: OPINION X GENDER
             35              25
             25              15
             60              40
CHI-SQUARE VALUE = 0.17          D.F. = 2

Do you want to repeat with other data?   (Y/N)__
```

Fig. 10.10 Output of the CHI program for computing the chi-square value.

Therefore, we observe that with simple commands the user can obtain different types of output from the data files created in FoxPro. It is also possible to write simple programs to derive custom-designed output for statistical calculations.

REFERENCES

1. Richard A. Johnson (1995): *Miller and Freund's Probability and Statistics for Engineers*, Prentice-Hall of India, New Delhi.
2. Taxali, R.K. (1996): *FoxPro 2.5 Made Simple for DOS and Windows*, BPB Publications, New Delhi.

DO IT YOURSELF

10.1 Create a blank data file to enter the data corresponding to the names and addresses of your classmates.

10.2 Attempt to write simple FoxPro programs for performing t-test based on the summary values like the mean, standard deviation and the sample size. This programs helps when the actual data file is not available and the researcher wishes to perform the t-value based on summary statistics.

10.3 A FoxPro file contains the marks of 25 students in maths, statistics and computer science corresponding to the hall ticket numbers. We wish to find the total marks and rank the students according to decreasing order of marks. Write a simple FoxPro program to do this job.

10.4 Attempt to work out Problem 10.3 using the rank and percentile option in the Data Analysis Park of Excel.

10.5 Create a data file and fill it with 50 blank records using BLANK.PRG.

10.6 Create a data file and calculate the descriptive statistics.

10.7 Examine the frequency distribution of selected variables in the file TRIBAL.DBF.

10.8 Create your own directory and type the programs given in this chapter. Try to use them.

Trend Analysis and Related Tools in Excel

Apart from statistical functions Excel has several other functions, which are very helpful in data analysis. The mathematical, financial and engineering functions in the category of Paste functions are very useful for quick handling of queries. For an economist, the problem of forecasting or finding the growth values is a common problem. Excel has ready-to-use functions for this job. For a theoretical statistician, one problem is about generating random numbers. Excel generates the required number of random numbers from the desired distribution. Similarly, the financial and engineering functions have specific uses of Excel. In this chapter, we discuss some important modules.

11.1 FITTING A TREND LINE TO THE OBSERVED DATA

Analysis of trends shown by time series data is an important aspect of statistical inference. Research problems in economics, commerce and management usually contain this type of problems. Business forecasting, marketing analysis, projections of share market and several related aspects belong to *trend analysis*. Decision makers often look at growth rates as indices of performance. Statistical methods help in providing a scientific basis for the study of the growth in any phenomenon.

Now let us understand some aspects of fitting a mathematical equation (called *model*) to the observed data. Reconsider the data in the file FOOD.DBF and construct a line chart showing the production values over different years.

Now right-click on this line and choose the option Trend Line from the resulting menu. A dialogue box appears as shown in Figure 11.1 for selecting the model and fit it to the observed line. By looking at the shape of the line shown by the data, we can select the suitable type and click on it. The options available for choosing the model are also shown as in Figure 11.1. The following are the various types of models available.

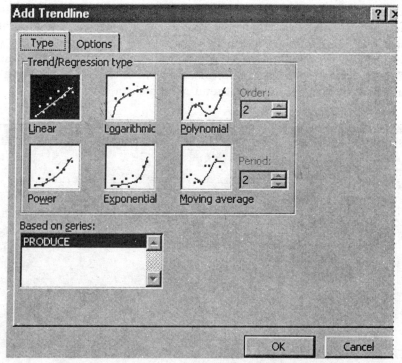

Fig. 11.1 Options for selecting the trend line model.

Linear model. This is used, when the data shows a fairly constant increase or decrease.

Logarithmic model. This model is used when the response increases or decreases at a constant ratio instead of a constant difference. Many economic time-series data are explained by this model.

Polynomial model. This is used, when the data has the shape of a curve with one or more changes. If the data shows an increasing trend followed by a decreasing trend, we choose the polynomial of degree 2. If there are two or more changes, we have to choose a higher degree like 3, 4 or 5. The menu provides for a degree up to 6.

Power model and exponential models. These are normally used for economic time series data for which the growth is either (i) initially slow and then increases or (ii) initially high and then decreases.

Moving averages. This is also called a *running average* and is used to obtain forecasts with the most recent 5 or 6 values to work out the mean. This is a useful method for following the track of the original data when there are frequent changes along with an overall increasing or decreasing linear trend.

Let us select the linear (straight line) model for explaining the observed trend with a right click on OK. After selecting the type of the model, we can choose different options regarding the adequacy of the fit in terms of R^2 value. We can also opt for forecasting the values for future periods. To do this, we have to click on the Options button. We then get another menu as shown in Figure 11.2.

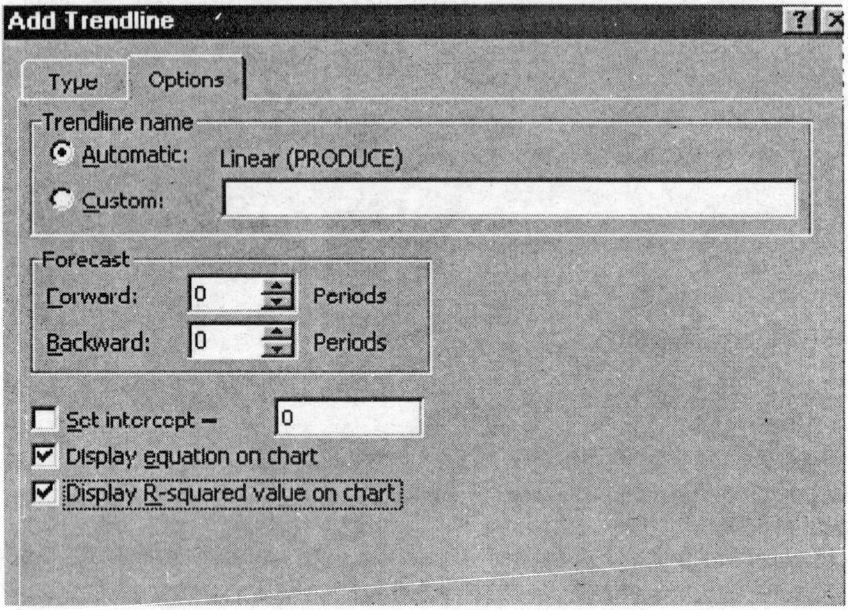

Fig. 11.2 Options for trend line display.

After choosing the type of the model to be fitted, we can look at the options. We then select the options *display equation on chart* and *display R-squared value*. If we select them, statistical model fitted to the data appears on the chart along with the R^2 value. A higher value or R^2 indicates a better fit while a smaller value indicates lack of fit of the chosen curve to the observed line.

These options are used in the following example.

EXAMPLE 11.1 Let us consider the data for the PRODUCTION figures of food grains as shown in Table 3.2. We choose a linear model (straight-line mode) to represent this data and observe its statistical properties. This curve is shown as in Figure 11.3.

Analysis It follows that the linear model fitted to the observed trend in production is given by the equation $Y = 3.2703 * X + 94.178$ with $R^2 = 0.8002$. Here, Y denotes production and X the year. The R^2 value shows that about 80% of the variation in production is explained by time factor (year) through this linear model. ∎

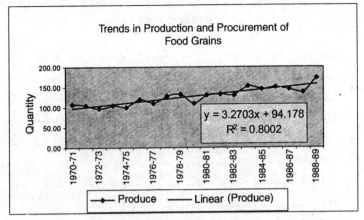

Fig. 11.3 A linear trend line fitted to the food grain data.

The researcher can report the R^2 value whenever an equation is fitted to the data. The fitted equation can now be used for forecasting the future values of production assuming that the same trend continues.

We can also make a forecast of the future values from the graph itself by selecting the options as shown in Example 11.2.

EXAMPLE 11.2 The price index of a commodity in different months of a year is shown below:

Month	Index	Month	Index
Jan	112	Jul	120
Feb	115	Aug	118
Mar	114	Sep	120
Apr	117	Oct	122
May	117	Nov	120
Jun	118	Dec	123

Let us fit a linear model using Excel graphs, and examine the trend line.

Analysis By selecting the line chart or the scatter diagram we can plot this data as done in Figure 11.3. We can now select the number of periods into which we wish to forecast forward. By default this period is set as 0. Let us select 2 future periods to be forecasted. Then Excel shows a trend line into future periods as shown in Figure 11.4.

This trend line is only an indication of the future values. The equation displayed on the graph shows a high value of R^2, which means that

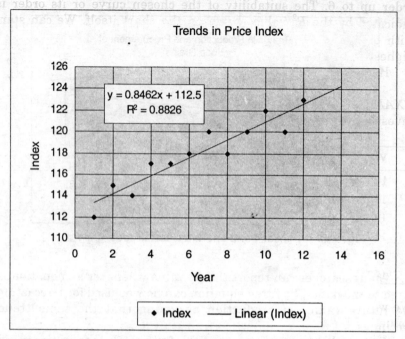

Fig. 11.4 Fitting of a trend line to the time series.

the linear equation is a good fit. If we consider the 12th month, we find that the trend line extends beyond it up to 14th month. This is because we have selected two periods for forecasting. However, the actual value of the forecast cannot be obtained from this chart. There is a separate procedure to obtain the forecast, which we shall discuss in the latter sections. We have to only understand that the graph is an indication of the possible future trend and helps as a 'first-aid' for further analysis. ■

11.2 POLYNOMIAL TRENDS

When the time series data cannot show a straight-line trend, we have to consider one suitable model from the available non-linear forms. The immediate alternative is a *quadratic* trend, which is also known as a *parabolic* trend. If the data has an increasing trend followed by a decreasing trend or vice versa, we consider a *parabolic* trend. Excel has a module in the trend-line options for a type of curve called Polynomial Curve. It is characterized by a parameter called the Degree, which indicates the highest-order term in the mathematical model. A polynomial of order 2 is called a *quadratic* curve or a *parabola*.

When the data shows both ups and downs we can consider a polynomial of order higher than 2. Excel supports a polynomial of order up to 6. The suitability of the chosen curve or its order is indicated by the R^2 value shown on the chart itself. We can start with order 2 and try fitting higher order curves until we get the highest R^2 value.

Here is an example of a non-linear trend in a time series data.

EXAMPLE 11.3 The percentage of dividend received against the investments in a company has been reported as follows.

Year	Dividend	Year	Dividend
1985	2.90	1991	2.80
1986	3.20	1992	2.20
1987	3.00	1993	2.00
1988	3.15	1994	1.50
1989	3.20	1995	1.40
1990	2.70	1996	1.00

We wish to know the trend in dividend and comment on the future values.

Analysis First we plot the given data as scatter diagram. We observe a curvilinear trend with the shape of an inverted bowl (called the concave shape). In order to fit a trend line, we right-click on the plotted line and choose from among the types the Polynomial type with degree 2. Then we get the trend line as shown in Figure 11.5.

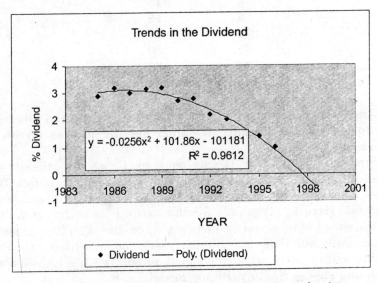

Fig. 11.5 Polynomial trend fitted to the scatter of points.

This graph shows a declining trend in the dividend in different years. Until 1981, the trend is more or less stable but a decline started. The projected line shows a negative value for the year 1998. This information helps in choosing a correct model for forecasting the future values.

For the same data, if we fit a linear trend model we find a lower value of R^2. Hence, while inspecting the data we have to judge the type of the model and then proceed to fit it. ∎

11.3 LOGARITHMIC, POWER AND EXPONENTIAL TRENDS

Economic and financial time series data sometimes shows a non-linear trend that can be better explained by a logarithmic curve, a power curve or an exponential curve. Excel has separate options for these two types of curves (see Figure 11.1). Here is an example of a case in which logarithmic curve is found suitable to the data.

EXAMPLE 11.4 Consider the following data on X and Y values, where the X values are increasing at a constant ratio instead of a constant difference. If we look at the scatter diagram of this data shown in Figure 11.6, we find that the increase in Y is at a faster rate at the beginning of the series and afterwards the rate of increase has become more or less a constant.

X	Y
2	10
4	15
8	19
16	26
32	30
64	34
128	39

Analysis We can choose a logarithmic curve and add this trend line to the observed trend. We then get the plot as shown in Figure 11.6.

Since the value of R^2 is very close to '1', we can consider the model to be a good fit to the observed scatter of the points. The model fitted is of the form $Y = a * \ln(x) + b$, where a and b are constants. Here $\ln(x)$ stands for the natural logarithm of X. The application of this model is justified when the X values increase geometrically but Y increases linearly. Instead of verifying the data for this nature, it is more convenient to observe the trend on the graph and choose the logarithmic model. ∎

Consider Example 11.5.

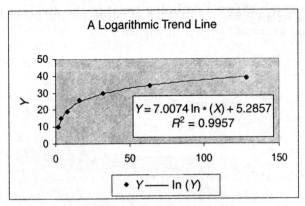

Fig. 11.6 A logarithmic curve fitted to the scatter of points.

EXAMPLE 11.5 Suppose we interchange the values of X and Y in Example 11.4 and get the following data.

Y	X
2	10
4	15
8	19
16	26
32	30
64	34
128	39

Here the values of X are increasing according to a linear scale and the Y values are increasing geometrically. We wish to fit a suitable model to this data.

Analysis The scatter plot suggests a power curve as a good fit to the data. The actual data and the fitted curve are shown in Figure 11.7.

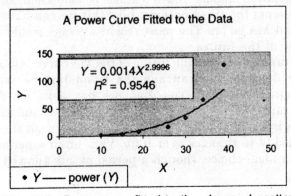

Fig. 11.7 Power curve fitted to the observed scatter.

Suppose we fit an exponential trend curve for the same data. Then we get the graph as shown in Figure 11.8.

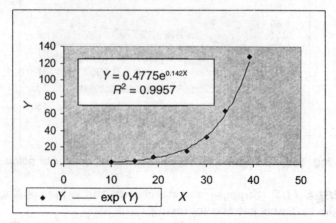

Fig. 11.8 Exponential curve fitted to the observed scatter.

It follows that the exponential curve has a better fit than the power curve for the same data, since the R^2 value is higher in the former case. ■

Thus, we can fit alternative models to the same data and choose the best fit based on the R^2 value.

11.4 MOVING AVERAGES

When the time-series data contains several changes—increasing and decreasing—one method of describing the data is *moving averages* (abbreviated as MA). It is also known as a *running* average and based on a few periods in the time series history. For instance a MA of period 3 is the average of first, second and third values, then second, third and fourth values and so on. By this method, the internal variation among the 3 values is smoothed and a single mean represents the three periods. Like this we can have a 5-period MA, 6-period MA so on. The most recent average would be used as the forecast of the future.

Excel graphics support the fitting of the MA curve. After preparing the scatter diagram, we can add the trend line by selecting the option Moving Averages. The dialogue box asks for the period, which we can give as 2 or 3 depending on our judgement. One simple rule is to take a higher MA period if we wish the trend line to react quickly to changes over time. Very often a period of 3 or 5 would be an ideal choice, though a period of 2 is allowed by default in Excel.

Consider the following example in which we find several changes in the data over time. So, we can fit an MA curve.

EXAMPLE 11.6 The quarterly sales of a soft drink (in thousand crates), bottled by a company in different years are given as in Table 11.1.*

Table 11.1 Quarterly Sales of a Soft Drink

Quarter	Sales	Quarter	Sales
91-1	41.8	93-3	65.3
91-2	82.3	93-4	45.2
91-3	60.7	94-1	70.7
91-4	43.8	94-2	101.3
92-1	55.7	94-3	110.6
92-2	93.5	94-4	74.2
92-3	94	95-1	90.1
92-4	53.1	95-2	100.6
93-1	75.6	95-3	110.4
93-2	93.6	95-4	78.6

*Data courtesy: Lavanya and Rajani, Project report, Department of Statistics, S.V. University, 1996.

The quarters are designated by year and quarter number. Thus 91-3 means the third quarter of the year 1991. The actual data has been slightly modified to alter the trend for the sake of demonstration.

Analysis The plot of this data shows a trend with seasonal variations. If we choose the moving average model to represent the trend, we get the type of trend line as shown in Figure 11.9.

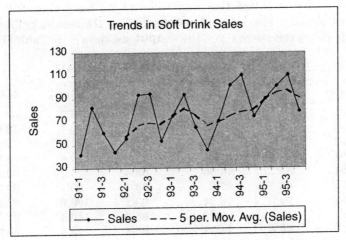

Fig. 11.9 Moving average trend line fitted to the sales data.

The moving average line is superimposed on the actual trend. It differs a lot from the original trend. It is to be understood only as the general trend, after accounting for the internal variation of the 5-period MA that has been used in the model. This curve is not a regression model and hence the equation and the R^2 cannot be displayed on the chart.

We can also carry out this analysis from the Data Analysis Pak. We can select the chart options in order to view the graphic output. The dialogue box for MA analysis appears as shown in Figure 11.10.

The input range is only the actual sales data. As done in the graph we need not specify the time axis as input. Excel automatically identifies the data points as 1, 2, 3 and so on. The next important factor is the column (or row) to show the output. The point to be noticed is that for every sales value, Excel provides a forecast as output. So, we have used column C for this purpose as shown in Figure 11.10.

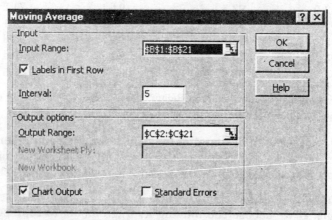

Fig. 11.10 Dialogue box for MA analysis.

The MA period is called the 'interval' and we have chosen it as 5. The chart output is also selected in order to view the results graphically. With all this exercise we get the output as shown in Table 11.2.

Table 11.2 MA Values for the Sales Data

Quarter	Sales	MA	Quarter	Sales	MA
91-1	41.8	#N/A	93-4	45.2	66.56
91-2	82.3	#N/A	94-1	70.7	70.08
91-3	60.7	#N/A	94-2	101.3	75.22
91-4	43.8	#N/A	94-3	110.6	78.62
92-1	55.7	56.86	94-4	74.2	80.4
92-2	93.5	67.2	95-1	90.1	89.38
92-3	94	69.54	95-2	100.6	95.36
92-4	53.1	68.02	95-3	110.4	97.18
93-1	75.6	74.38	95-4	78.6	90.78
93-2	93.6	81.96			
93-3	65.3	76.32			

The first four values under the column MA are shown as #NA, which means that the MA, forecast will be available only from the 5th period onwards.

Then the next 5 periods are averaged and the value is written against the 6th period. This method is continued till the last period. The last MA will be forecast for the first quarter of 1996. When the actual sales is realized in the next period, the forecast can be updated (Figure 11.11).

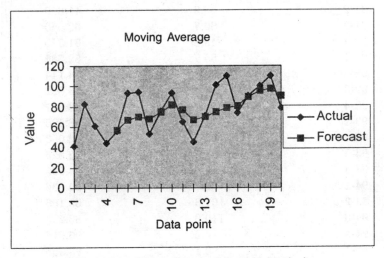

Fig. 11.11 Chart output of the MA analysis.

As an option, we can select the standard errors of the forecasts to be shown in the data output.

11.5 EXPONENTIAL SMOOTHING

This is another method of describing a time series and providing a forecast for the future. This method is like that of moving averages except that instead of specifying the discrete MA period (like 3, 4 or 5) we specify a factor called the *damping factor* or *smoothing factor*. Excel has a separate module in the Data Analysis Pak to carry out this analysis.

In the following example, exponential smoothing is used instead of MA method for the data given in Example 11.6.

EXAMPLE 11.7 Reconsider the Excel worksheet for the data given in Example 11.6 and open Data Analysis Pak.

Analysis Select the item Exponential Smoothing and click on the Data Range. After selecting the Input Data, we have to specify the damping factor α, which is usually taken between 0 and 0.2. A higher value of this factor leads to a quick response to changes over time but results in erratic projections. We have however taken $\alpha = 0.5$ and the output shown in Table 11.3 is obtained.

Table 11.3 Exponentially Smoothed Values for the Sales Data

Quarter	Sales	Exponential forecast
91-1	41.8	#N/A
91-2	82.3	41.800
91-3	60.7	62.050
91-4	43.8	61.375
92-1	55.7	52.588
92-2	93.5	54.144
92-3	94.0	73.822
92-4	53.1	83.911
93-1	75.6	68.505
93-2	93.6	72.053
93-3	65.3	82.826
93-4	45.2	74.063
94-1	70.7	59.632
94-2	101.3	65.166
94-3	110.6	83.233
94-4	74.2	96.916
95-1	90.1	85.558
95-2	100.6	87.829
95-3	110.4	94.215
95-4	78.6	102.307

The chart output for this forecasting method appears as shown in Figure 11.12.

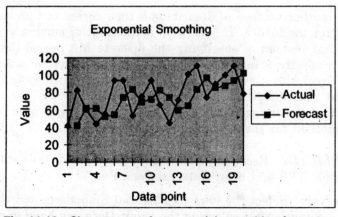

Fig. 11.12 Chart output of exponential smoothing forecast.

11.6 LINEAR AND COMPOUND GROWTH VALUES

Excel has some built-in functions for computing the growth values of a time series data using either a linear model or an exponential model. We can select the following Paste functions to forecast future values. The operating procedure and the type of the output we get are discussed below.

11.6.1 The Forecast Function

This is a built-in function used to obtain a future value of a series based on a linear trend model. This function gives a quick result in contrast to the trend line method applied to the scatter diagram. In the list of statistical functions, we find a Paste function called Forecast. If we click on it we get a dialogue box and the inputs into it gives the output. Consider Example 11.8.

EXAMPLE 11.8 The monthly sale of cars of a popular company in a city is given below.

Month	Cars sold	Month	Cars sold
1	45	7	47
2	52	8	51
3	41	9	58
4	36	10	47
5	42	11	52
6	49	12	50

We wish to forecast the sales for the 13th month.

Analysis To do this, we first enter the data in a worksheet as two columns. At the end of the data, we type 13 under column Month. Against the month 13 we have to place the cursor and select the Paste functions. In the category of statistical functions, we find the function Forecast. The dialogue box, which is self-explanatory, appears as shown in Figure 11.13.

In this box, the notation Y indicates the sales and X indicates the month. We can forecast the value for the 13th month by filling the inputs in the dialogue box. It is important to note that Forecast gives prediction assuming that the trend in the data is *Linear* (like a straight line).

After entering these details, we get the output against the 13th month as 52.77 or approximately 53 cars. We can copy this function to further periods 14, 15, ..., 20 and get the forecast values.

A related function to Forecast is the Linest function. It gives the *line estimate* of the series to mean the *slope* of the straight line that

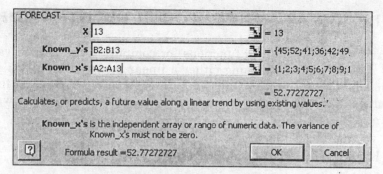

Fig. 11.13 Dialogue box for the FORECAST function.

represents the data. To use this function, we have to again select it from the Paste functions menu. A dialogue box appears for inputs and we get the value of the slope for the linear model. This is very useful in calculating *linear growth rate* (LGR). In this case, the Linest is seen to be 0.8112.

The output of Forecast function is shown in Table 11.4. The Linest is nothing but the regression coefficient in the linear model that fits to the data. If we fit a linear trend model, we get

$$Y = 0.8112X + 42.227$$

By putting $X = 13$, we get $Y = 52.7726$, which is the same as the Forecast value shown in Figure 11.13. We note that the coefficient of X in this model is the Linest value.

It is also customary to express the linear growth rate (LGR) as a percentage given by

$$LGR = \frac{100 * Linest}{Average\ sales}$$

In this case, the LGR rate will be 1.71% per month.

Table 11.4 Future Values Obtained with Forecast Function

Given Series		Forecast Values	
Month	Cars sold	Month	Cars sold
1	45		
2	52	13	52.77
3	41	14	54.00
4	36	15	56.77
5	42	16	58.03
6	49	17	57.95
7	47	18	50.10
8	51	19	59.06
9	58	20	59.51
10	47		
11	52		
12	50		

We can as well use another function called Trend in place of Forecast to get the future values along a straight line. If we wish the forecast for further periods we can simply copy the cell result to other cells down in that column.

11.6.2 The Growth Function

When the time series data shows a non-linear trend, usually a Growth model is used to predict the future values. Many of the economic time series data are characterized by a *compound growth* based on a model of the type $Y = a * b^X$. The forecast is obtained by selecting the Growth function and filling the inputs in the dialogue box, which is similar to the Forecast dialogue box. Excel shows this model as $Y = b * m^{\wedge}X$. (See Help before applying function.)

A related function to Growth is known as Logest to mean the estimate of the *logarithmic growth*. The Logest is similar to the Linest function. It gives the value of the constant 'm' in the growth model. We can use this to estimate the compound growth rate (CGR) given by

$$\text{CGR} = (\text{Logest} - 1) * 100$$

In this case we get CGR = 1.778%.

When we use Growth functions on the data of Example 11.8, we get the output shown in Table 11.5.

Table 11.5 Future Values Obtained with Growth Function

Given series		Future values	
Month	Cars sold	Month	Cars sold
1	45	13	52.87
2	52	14	53.18
3	41	15	54.77
4	36	16	55.74
5	42	17	56.73
6	49	18	57.74
7	47	19	58.77
8	51	20	59.82
9	58		
10	47		
11	52		
12	50		

A comparison of the predicted value for the 13th period based on Forecast and the Growth functions did not differ much. The difference between the LGR and the CGR is also not much. One reason for this is the scatter diagram for this data shows a trend as in Figure 11.14, which has an overall increasing trend with some variations.

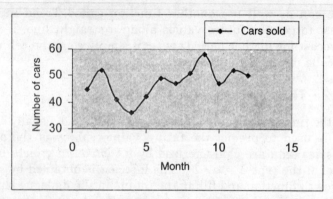

Fig. 11.14 Trend in the sale of car, which at most, recommends a linear fit with variations.

Hence the researcher has to use his own judgement regarding which model to use for forecasting, depending on the scatter or line diagram of the actual data.

11.7 FINANCIAL FUNCTIONS AND RELATED TOOLS IN EXCEL

There are several helpful modules for the common user as well as the researcher. The financial functions, Date and Time functions have interesting applications. For a theoretical statistician, there are special tools and generating data. We will briefly discuss some functions of that type in the following sections. For more details, one may use Excel online help and learn the details.

11.7.1 Simple Financial Functions

Excel supports several financial functions that are useful to statisticians or researchers in the field of commerce or management. One of the Paste functions is designated as Financial functions. By clicking on, we get a list of available financial functions. If we select any function the corresponding dialogue box appears on the screen and the input can be fed in the entry boxes. This gives the corresponding financial parameter. For example, if we click the FV (future value) the dialogue box appears as in Figure 11.15. It gives the future value of an investment at the specified rate of interest. The dialogue box asks for several inputs and explains the meaning of each of the input.

For sample values, which are chosen arbitrarily, the output appears as shown in the above figure.

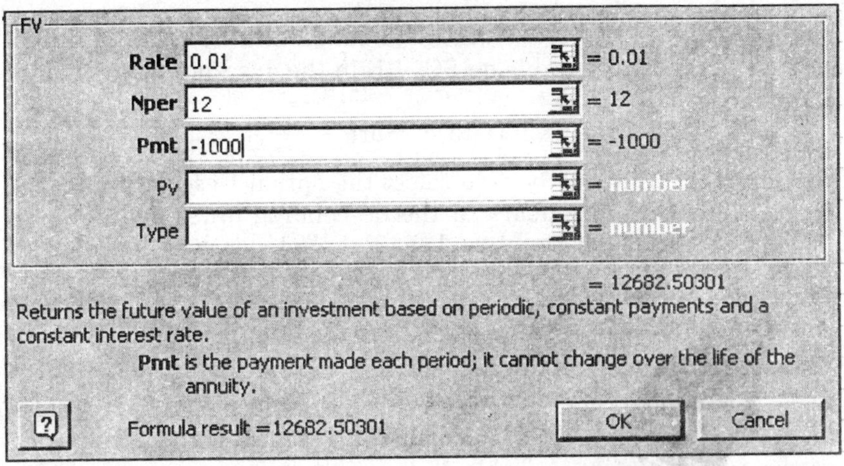

Fig. 11.15 Dialogue box for computing the Future Value of a payment.

11.7.2 Ranking of Data

Excel has a module to perform the job of ranking a list of values. This is a common requirement for evaluating the performance of students or other experimental subjects based on their performance expressed as a score. For instance, we have a list of 10 students with details on hall ticket: name, number and the test score. We can rank them by selecting the item 'Rank and Percentile' from the Data Analysis Park. The use of this module is discussed in Example 11.9.

EXAMPLE 11.9 The following are the test scores of 10 students. We are required to rank the students according to their scores. The top ranker should get the highest score and students with equal ranks should be given the same rank. The output should be the name, the hall ticket number and score of the student along with the rank. The data written in the alphabetical order is shown as below:

NAME	HTNO	SCORE
JOSEPH. N	28117	77
JYOTHI. V	28115	68
KARTHIK. M	28112	82
LATHA. J	28114	79
NEEHARIKA. G	28119	79
RAFI. J	28111	82
RANI. V	28118	81
RANJITHA. L	28120	53
SRIKANTH. L	28113	58
VICTOR. B	28116	82

Analysis In order to rank these candidates according to their score we first sort this table on SCORE in the descending order. The command is

Data ▸ Sort

Then from the dialogue box we select the option Descending Order. Then the entire list appears in the descending order of the score. Now, let us select the Data Analysis Pak and select the item Rank and Percentile. The dialogue box asks for the input range. We give the input range as the column containing the score.

The box also asks for the output range. We can either choose the New Worksheet or in the Same Worksheet with different columns. Let us select the option of Same Worksheet and click on the column which is to the right of SCORE column. With all these options, we get the output shown in Table 11.6.

Table 11.6 Students' Data After Ranking

NAME	HTNO	SCORE	Point	Rank	Per cent
RAFI. J	28111	82	1	1	77.70
KARTHIK. M	28112	82	2	1	77.70
VICTOR. B	28116	82	3	1	77.70
RANI. V	28118	81	4	4	66.60
LATHA. J	28114	79	5	5	44.40
NEEHARIKA. G	28119	79	6	5	44.40
JOSEPH.N	28117	77	7	7	33.30
JYOTHI. V	28115	68	8	8	22.20
SRIKANTH. L	28113	58	9	9	11.10
RANJITHA. L	28120	53	10	10	00.00

We find that the first three candidates who have tied ranks have been given rank 1. If there is any other criterion for ranking like date of birth we have to first sort the data on both SCORE and the DATE OF BIRTH and then repeat the above procedure. This resolves the tied ranks to some extent. One has to define rules for resolving tied ranks before using the ranking procedure.

11.7.3 Random Number Generation

There is a special type of decision making tool called Simulation. When we wish to study some aspects of a real system like the future cash-flow of a super market, one method is to use a mathematical model to describe the operations. The model is then solved for the values of the decision variables such that they optimize some objective function like profit maximization or minimization of losses. Such models tend to be very complicated when the system is complex. In such cases a second approach called Simulation is adopted.

In simple words, it is a method of *imitating* the real system with artificial data that can closely represent the actual transactions of the real system. Such a data is generated with the help of numbers called *random numbers*.

Every system, which has an element of uncertainty, can be described by a theoretical distribution like binomial, Poisson, normal and so on. We may be required to generate artificial data that could possibly be treated as a sample from one of these distributions.

Excel provides a direct module in the Data Analysis Pak to generate random numbers from a selected distribution. By clicking on this item, a dialogue box appears as shown in Figure 11.16 and by filling the input details in it, we get the required samples. It asks for the number of variables on which the random numbers should be generated. We have chosen this number as 10. Hence we get 10 columns of data. In each column we need 15 random numbers. So, we specify the number of random numbers as 15.

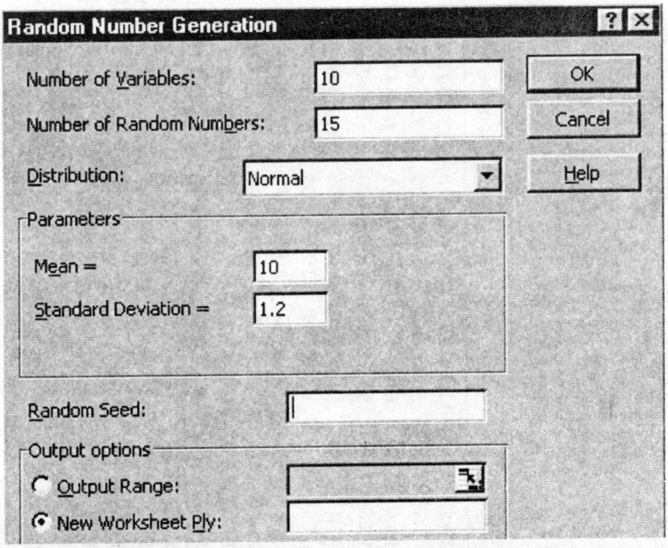

Fig. 11.16 Dialogue box for generating random numbers from normal distribution.

The next important aspect is the *distribution*. By clicking in that box we get a list of possible distributions from which we have to generate the random numbers. Here, we select the normal distribution with mean 10.00 and standard deviation 1.2.

The next item is called the *random number seed*. It is an optional value, which we may specify so that the random numbers are generated around this value. We can skip if we do not have a seed to specify.

In Table 11.7, ten samples each containing 15 values from a normal distribution with mean 10.00 and standard deviation 1.2 is shown.

Table 11.7 Ten Samples of Random Numbers Generated from a Normal Distribution

Sample-1	Sample-2	Sample-3	Sample-4	Sample-5	Sample-6	Sample-7	Sample-8	Sample-9	Sample-10
9.639721	8.466780	10.293110	11.531770	11.438020	12.079760	7.379695	9.718983	11.314030	8.695959
9.171755	7.971481	7.783707	8.826845	9.071792	7.458483	9.318490	9.515143	10.161820	9.561408
9.607611	9.555711	11.611170	9.897659	9.776611	9.384151	12.366650	11.038810	12.850790	9.214112
11.993750	8.065123	10.646740	11.082630	12.302700	9.898580	9.371446	10.810170	9.542411	10.909130
8.266976	8.983315	8.174115	9.564548	9.961025	10.033740	9.612741	12.633400	7.909021	9.116228
6.906903	11.737200	8.464284	9.215704	10.909260	10.560050	11.049530	10.714890	8.353780	8.661114
10.832790	10.387160	8.872195	9.710863	10.157840	10.669360	10.166460	8.906846	12.261820	10.584640
10.086690	10.995810	11.034410	9.236162	8.892170	11.333430	8.558586	8.129329	10.853590	10.766090
12.646830	11.732510	11.564680	10.135550	10.002340	10.544440	9.969382	8.734390	7.870233	10.994000
10.533070	10.741490	10.256170	8.767683	11.485830	9.626544	8.992094	9.014646	9.485209	9.455966
9.371446	11.019320	10.615850	9.270035	11.565980	7.886876	10.660690	9.860474	10.050090	9.215135
9.340487	11.019180	10.963650	10.547700	10.829640	11.956750	10.364700	10.706790	12.223400	9.597332
11.245930	10.172210	11.369250	9.822041	9.064220	11.291190	9.318140	10.640690	10.654930	9.620276
9.471081	8.360809	12.391550	9.320215	10.103350	9.718887	13.402440	11.502370	11.057090	11.598600
10.225450	10.650350	9.700456	8.534695	11.520020	9.653335	8.432950	10.917100	10.941140	10.513790

Excel does not provide headings for the samples by itself. We have to provide headings after pasting that table in MS-Word for easy understanding and apply autoformat for the table.

If we again work out the mean and standard deviation of all these 10 samples taken as a single sample of 150 values, we get a mean of 10.08948 and a standard deviation 1.25362.

These values are close to the parameters (mean = 10 and S.D. = 1.2) which we have taken for the normal distribution. Thus, we can treat 10 samples as if they have been drawn at random from a normal population with the specified parameters.

Simulation of data has a large number of applications in operations research and other management studies.

11.7.4 Creation of Statistical Tables

One classical approach is to refer to the table of critical values from the selected distribution—normal distribution, student's t-distribution, F-distribution, etc. In Excel, there is no need for reading the critical value because the Data Analysis Pak automatically gives the critical value at 0.05 and 0.01 levels. In fact, we can draw valid inferences using the p-value of a test instead of comparing the calculated value with the critical value.

However, if required, the researcher can prepare tables of critical values from selected distributions with the help of the Paste functions. By selecting the normal distribution function Normdist, we can create the cumulative normal distribution values for different values of x. The dialogue box asks for the mean and standard deviation as inputs. It also asks to confirm whether the required value is from cumulative distribution or not. If we type TRUE we get the cumulative value. Table 11.8 is generated from Excel by this method.

Here we have used mean = 0 and standard deviation = 1 so that each value becomes the area under the cumulative standard normal distribution from $-\infty$ to z, where z is the given value. This gives the value in one cell. We have to copy this formula to other cells of the column. We have also used a simple formula to compute the probability by combining the value given along the rows with the header value. For instance, the cumulative probability for the standard normal distribution, up to $z = 1.53$ can be read as 0.9370 from this table.

We can similarly create tables of other distributions with the help of the Paste functions.

Conclusively, we can say that Excel functions can be utilized for data analysis in several ways. Depending on the need we can select the suitable function and obtain the results. We may use simple programming steps or formulas to perform repeated calculations.

Table 11.8 Cumulative Standard Normal Distribution Values from $-\infty$ to z

z	0.00	0.01	0.02	0.03	0.04	0.05	0.06	0.07	0.08	0.09
0.00	0.5000	0.5040	0.5080	0.5120	0.5160	0.5199	0.5239	0.5279	0.5319	0.5359
0.10	0.5398	0.5438	0.5478	0.5517	0.5557	0.5596	0.5636	0.5675	0.5714	0.5753
0.20	0.5793	0.5832	0.5871	0.5910	0.5948	0.5987	0.6026	0.6064	0.6103	0.6141
0.30	0.6179	0.6217	0.6255	0.6293	0.6331	0.6368	0.6406	0.6443	0.6480	0.6517
0.40	0.6554	0.6591	0.6628	0.6664	0.6700	0.6736	0.6772	0.6808	0.6844	0.6879
0.50	0.6915	0.6950	0.6985	0.7019	0.7054	0.7088	0.7123	0.7157	0.7190	0.7224
0.60	0.7257	0.7291	0.7324	0.7357	0.7389	0.7422	0.7454	0.7486	0.7517	0.7549
0.70	0.7580	0.7611	0.7642	0.7673	0.7704	0.7734	0.7764	0.7794	0.7823	0.7852
0.80	0.7881	0.7910	0.7939	0.7967	0.7995	0.8023	0.8051	0.8078	0.8106	0.8133
0.90	0.8159	0.8186	0.8212	0.8238	0.8264	0.8289	0.8315	0.8340	0.8365	0.8389
1.00	0.8413	0.8438	0.8461	0.8485	0.8508	0.8531	0.8554	0.8577	0.8599	0.8621
1.10	0.8643	0.8665	0.8686	0.8708	0.8729	0.8749	0.8770	0.8790	0.8810	0.8830
1.20	0.8849	0.8869	0.8888	0.8907	0.8925	0.8944	0.8962	0.8980	0.8997	0.9015
1.30	0.9032	0.9049	0.9066	0.9082	0.9099	0.9115	0.9131	0.9147	0.9162	0.9177
1.40	0.9192	0.9207	0.9222	0.9236	0.9251	0.9265	0.9279	0.9292	0.9306	0.9319
1.50	0.9332	0.9345	0.9357	0.9370	0.9382	0.9394	0.9406	0.9418	0.9429	0.9441
1.60	0.9452	0.9463	0.9474	0.9484	0.9495	0.9505	0.9515	0.9525	0.9535	0.9545
1.70	0.9554	0.9564	0.9573	0.9582	0.9591	0.9599	0.9608	0.9616	0.9625	0.9633
1.80	0.9641	0.9649	0.9656	0.9664	0.9671	0.9678	0.9686	0.9693	0.9699	0.9706
1.90	0.9713	0.9719	0.9726	0.9732	0.9738	0.9744	0.9750	0.9756	0.9761	0.9767
2.00	0.9772	0.9778	0.9783	0.9788	0.9793	0.9798	0.9803	0.9808	0.9812	0.9817
2.10	0.9821	0.9826	0.9830	0.9834	0.9838	0.9842	0.9846	0.9850	0.9854	0.9857
2.20	0.9861	0.9864	0.9868	0.9871	0.9875	0.9878	0.9881	0.9884	0.9887	0.9890
2.30	0.9893	0.9896	0.9898	0.9901	0.9904	0.9906	0.9909	0.9911	0.9913	0.9916
2.40	0.9918	0.9920	0.9922	0.9925	0.9927	0.9929	0.9931	0.9932	0.9934	0.9936
2.50	0.9938	0.9940	0.9941	0.9943	0.9945	0.9946	0.9948	0.9949	0.9951	0.9952
2.60	0.9953	0.9955	0.9956	0.9957	0.9959	0.9960	0.9961	0.9962	0.9963	0.9964
2.70	0.9965	0.9966	0.9967	0.9968	0.9969	0.9970	0.9971	0.9972	0.9973	0.9974
2.80	0.9974	0.9975	0.9976	0.9977	0.9977	0.9978	0.9979	0.9979	0.9980	0.9981
2.90	0.9981	0.9982	0.9982	0.9983	0.9984	0.9984	0.9985	0.9985	0.9986	0.9986
3.00	0.9987	0.9987	0.9987	0.9988	0.9988	0.9989	0.9989	0.9989	0.9990	0.9990
3.10	0.9990	0.9991	0.9991	0.9991	0.9992	0.9992	0.9992	0.9992	0.9993	0.9993
3.20	0.9993	0.9993	0.9994	0.9994	0.9994	0.9994	0.9994	0.9995	0.9995	0.9995
3.30	0.9995	0.9995	0.9995	0.9996	0.9996	0.9996	0.9996	0.9996	0.9996	0.9997
3.40	0.9997	0.9997	0.9997	0.9997	0.9997	0.9997	0.9997	0.9997	0.9997	0.9998
3.50	0.9998	0.9998	0.9998	0.9998	0.9998	0.9998	0.9998	0.9998	0.9998	0.9998
3.60	0.9998	0.9998	0.9999	0.9999	0.9999	0.9999	0.9999	0.9999	0.9999	0.9999
3.70	0.9999	0.9999	0.9999	0.9999	0.9999	0.9999	0.9999	0.9999	0.9999	0.9999
3.80	0.9999	0.9999	0.9999	0.9999	0.9999	0.9999	0.9999	0.9999	0.9999	0.9999
3.90	1.0000	1.0000	1.0000	1.0000	1.0000	1.0000	1.0000	1.0000	1.0000	1.0000
4.00	1.0000	1.0000	1.0000	1.0000	1.0000	1.0000	1.0000	1.0000	1.0000	1.0000

REFERENCES

1. Alden, R. and Thomas Chester: *Mastering Excel* 97, BPB Publications, New Delhi.
2. Microsoft-Excel online help.

DO IT YOURSELF

11.1 Reconsider the data given in Example 11.2. Try to fit different possible curves and compare the corresponding R^2 values. Could you identify the best-fit curve?

11.2 Consider the data given in Example 11.6 and fit a polynomial curve.

11.3 The statistics of violence in Kashmir during 1990 to 2000 has been reported as follows. The figures indicate the toll in different years of civilians, militants and the security personnels.

Year	Civilians	Militants	Security personnels
1990	461	550	154
1991	382	844	168
1992	634	819	177
1993	747	1,310	195
1994	820	1,596	198
1995	103	133	234
1996	1,336	1,209	185
1997	938	1,075	186
1998	867	999	232
1999	821	1,083	356
2000*	143	183	68

Source: *The Hindu*, Sunday, April 2, 2000, p. 14.
*Till March 2000.

For each category of data, plot this data as a line graph. Obtain the best-fit curve and comment on the general trend.

11.4 Prepare a Word document with the list of financial functions available in Excel as a table. It should contain the name of the function, the input requirements, the type of output and convenient short cut for easy identity.

11.5 Look at the Data Analysis Pak and examine the aspects given in SAMPLING. It helps in taking a sample of data records from a file having a large number of values or records. Try this function with the file TRIBAL.DBF. or TRIBAL.XLS and get a random sample of 10 BMI values.

11.6 Generate a random sample of 20 values from a Poisson distribution with mean 5.

Software for Higher Statistical Analyses 12

Apart from the basic statistical tools provided by Excel, the researcher may need several advanced methods for data analysis. With the help of Internet, we can find out the availability of several statistical software packages. Some of them are free of cost. Though the objective of this book is only to motivate the researcher to utilize the commonly available software, we briefly mention the sources where advanced statistical tools can be found.

12.1 SPSS AND ITS APPLICATIONS

The Statistical Package for Social Sciences (SPSS) is a popular software package for statistical calculations. Its version 7.5 works in Windows 95/98 environment and supports linking to other software like MS-Office. The following are the main features of SPSS:

1. It can read dBASE files directly.
2. It can filter the data and perform analysis only in the selected cases.
3. Its output can be pasted in other documents in Windows.
4. It supports a large number of statistical graphs.
5. The help module gives quick reference to statistical concepts with a brief description. So, it is like a ready-reckoner.

SPSS opening screen appears as shown in Figure 12.1. It contains the main menu with different items on the top of the screen. The screen looks like a worksheet of Excel. Data can be entered directly in the cells. We can also read data from an existing file like a FoxPro file. When we click on the item Statistics we get a pull-down menu with different modules related to statistical analysis. We discuss some of them briefly.

12.1.1 Summary Statistics

Given a data file either in FoxPro or in SPSS, we can work out the

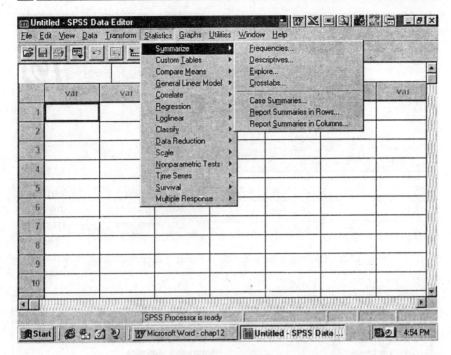

Fig. 12.1 SPSS opening screen.

statistical aspects of the data. *Summary statistics* is a common requirement, which involves the calculation of frequencies, descriptive statistics, preparation of tables and so on. Consider Example 12.1.

EXAMPLE 12.1 Let us open the file TRIBAL.DBF in SPSS. When we select the item *Descriptives* under the module summarize we get a dialogue box as shown in Figure 12.2 to select variables for computing the descriptive statistics. The list of all the variables in the file will be displayed in the box along with a provision for output options. If we click the item Options, we get another menu from which we can select the output like Mean, Variance, etc., and click on *Continue*. The control then goes back to Descriptives menu box in which we have to click OK. SPSS then starts working on these variables and the output appears as shown in Figure 12.3.

We have selected the statistics Mean and Standard Deviation only for illustration. When we select several statistics the output goes to more than one page. The Print Preview option is useful to view the output before taking the hard copy of the output.

232 STATISTICS MADE SIMPLE—DO IT YOURSELF ON PC

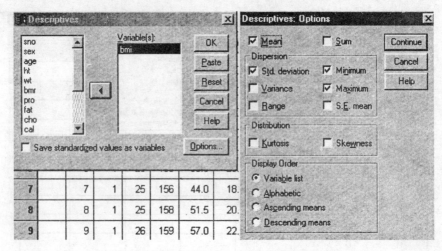

Fig. 12.2 Dialogue box for computing descriptive statistics.

Descriptives

Descriptive Statistics

	N	Mean		Std.
	Statistic	Statistic	Std. Error	Statistic
BMI	150	19.9299	.1732	2.1212
HT	150	155.36	.43	5.30
WT	150	48.157	.495	6.066
BMR	150	1324.536	13.287	162.735
PRO	150	48.967	.629	7.709
FAT	150	29.499	.667	8.167
CHO	150	418.981	6.534	80.031
Valid N (listwise)	150			

Fig. 12.3 SPSS output of descriptive statistics with selected options.

12.1.2 Cross Tabulations

This is another useful option for the preparation of two-way tables. From the raw data file, we can create Cross Tabulation and also perform chi-square test if necessary. Consider Example 12.2.

EXAMPLE 12.2 Reconsider the file TRIBAL.DBF. When we click on Summarize ▸ Crosstabs we get the corresponding menu box as shown in Figure 12.4. The entries in the box provide options for selecting the variables to be shown along the rows and columns.

SOFTWARE FOR HIGHER STATISTICAL ANALYSES

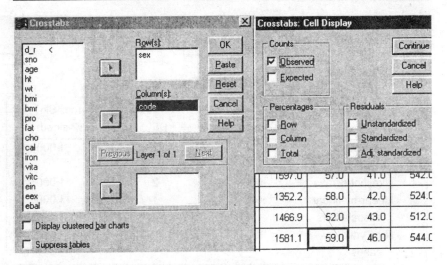

Fig. 12.4 Crosstabs dialogue box.

We can also select more than one variable along the rows or columns but one pair at a time appears to be good from the point of understanding the output.

We find an option Cells under which we can fix up what output should appear in the cells. Usually the Count of the observed cases is reported. If necessary, we can click on percentages according to our requirement. We may choose Row Percentages or Column Percentages or the Total Percentages.

The Crosstabs output between SEX and CODE appears as shown in Figure 12.5.

SEX-CODE Cross Tabulation

			CODE			
			1	2	3	Total
SEX	1	Count	25	25	25	75
		% within CODE	50.0%	50.0%	50.0%	50.0%
	2	Count	25	25	25	75
		% within CODE	50.0%	50.0%	50.0%	50.0%
Total		Count	50	50	50	150
		% within CODE	100%	100%	100%	100%

Fig. 12.5 Crosstabs of SEX versus CODE.

When we choose the Chi-square test to be performed on the cross tabulation, we get the output as shown in Figure 12.6.

Chi-square Tests

	Value	DF	Asymp. Sig. (2-sided)
Pearson chi-square	.000*	2	1.000
Likelihood ratio	.000	2	1.000
Linear-by-linear association	.000	1	1.000
N of valid cases	150		

* 0 cell (.0%) have expected count less than 5. The mininum expected count is 25.00

Fig. 12.6 Chi-square test performed on the cross tabulation.

If the table contains only 2 rows and 2 columns, SPSS conducts

(i) Fisher's exact test, provided the expected frequencies are less than 5 and there are no missing rows or columns.

(ii) For all other 2 × 2 tables, Yate's corrected chi-square test is computed.

(iii) For tables other than 2 × 2 size, Pearson's chi-square test and the likelihood ratio chi-square is computed.

The above criteria apply only when the variables along the rows and columns are categorical. However, if the variables happened to be quantitative, the chi-square and the linear-by-linear test yield the same result. ■

More details on chi-square test can be had from SPSS help.

12.2 INFERENTIAL TOOLS IN SPSS

SPSS offers a big range of statistical tests of significance, methods of correlation and regression and data reduction techniques. Each module is like a separate analysis tool. We can carry out the procedure by clicking on the corresponding module and following the instructions given in the dialogue box. Here is a brief summary of the tools available.

Tool/Module	Statistical procedures
1. Compare means	– Means – One-sample t-test – Independent sample t-test – Paired sample t-test – One-way ANOVA
2. General linear model	– Simple factorial – GLM–General factorial ... – GLM–Multivariate ... – GLM–Repeated measures ... – Variance components
3. Correlate	– Bivariate – Partial – Distance
4. Regression	– Linear – Curve estimation – Logistic – Probit – Non-linear – Weight estimate – 2-stage least squares
5. Log linear	– General – Logit – Model selection
6. Classify	– k-means cluster – Hierarchical cluster – Discriminant
7. Data reduction	– Factor (analysis) – Corresponding analysis – Optimal scaling – Reliability analysis – Multidimensional scaling
8. Non-parametric tests	– Chi-square – Binomial – Runs – 1, 2, k Sample K-S tests – 2 Related samples – k Related samples
9. Time series	– Exponential smoothing – Auto regression – ARIMA – XII ARIMA – Seasonal decomposition
10. Survival	– Lifetables – Kaplan–Meier – Cox regression – Cox w/Time-Dep Cov
11. Multiple response	– Define sets – Frequencies – Crosstabs

We find that some of these tools are available in Excel. However, the output given by SPSS contains several details regarding the statistical aspects of the findings. Here is an illustration for ANOVA.

EXAMPLE 12.3 Suppose we wish to perform a two-way analysis of variance of the shoot length (sl25) from the data given in the FoxPro file PLANT.DBF (discussed in Chapter 3). There are two factors namely TREATMENTS (1 to 4) and VARIETIES (1 to 3). We can now test for the main effects and the interaction.

Analysis We have to first open the file by selecting File ▸ Open. Then we select the file type as dBASE files we can open this file. Now, we have to select Statistics ▸ General Linear Model ▸ Simple Factorial. The dialogue box calls for fixing the dependent variable, the variable on which the ANOVA should be carried out. Here it is sl25. The factors are TREATMENT and VARIETIES. We have to enter the range for them. The next item is Options. We get another dialogue box in which we click on 2-Way under Interaction. The procedure is by default fixed as Unique. We can change to Hierarchical or Experimental, if needed.

The output of ANOVA appears as shown in Figure 12.7.

ANOVA[a,b]

			Unique method				
			Sum of squares	DF	Mean square	F	Sig.
SL25	Main effects	(Combined)	159.100	5	31.820	1.535	.197
		VARIETY	138.700	2	69.350	3.346	.044
		TREATMENT	20.400	3	6.800	.328	.805
	2-way interactions	VARIETY* TREATMENT	15.700	6	2.617	.126	.993
	Model		174.800	11	15.891	.767	.670
	Residual		994.800	48	20.725		
	Total		1169.600	59	19.824		

a. SL25 by VARIETY, TREATMENT.
b. All effects entered simultaneously.

Fig. 12.7 Two-way analysis of variance as reported by SPSS.

The output shows the actual field names so that the results can be understood easily. It contains two main effects and one interaction and the *F*-test as usual. ■

Regarding advanced statistical analysis tools, we have to look at non-linear regressions, time series analysis, fitting of probit and logit models, and so on. We may also require special tools like Factor Analysis, Cluster Analysis, Discriminant Analysis, and so on. All these aspects require some basic understanding of the method, and its applicability in the problem under study. (In view of the limitations of this book these advanced aspects are not discussed.)

12.3 GRAPHS AND OTHER ANALYSIS TOOLS IN SPSS

SPSS supports several statistical graphs like bar charts, pie charts, and so on. One interesting aspect is that, for a given raw data from a file, it can create a Histogram with automatic class intervals and also displays the mean and standard deviation of the data on the chart itself. It also provides special graphs to test for normality like Q-Q Plot, P-P Plot. Here is a simple histogram of BMI from the FoxPro file TRIBAL.DBF shown in Figure 12.8.

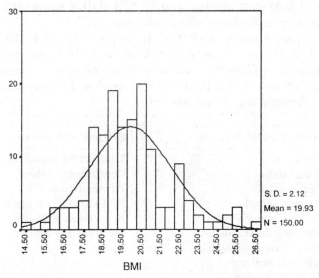

Fig. 12.8 Histogram of BMI prepared by SPSS.

We can also perform several routines required for data analysis, like generating random numbers, defining new variables, sorting data, recoding and transforming the variables, and so on.

The output of SPSS usually appears as Pivot Tables. As an options (available in Edit) we can prefer to have a SPSS Classical type of output which do not have tables. The output of SPSS can be cut and pasted to Word documents, Excel worksheet or a PowerPoint presentation.

When the quantum of work to be done on a file is large, like performing a large number of t-tests or ANOVA procedures, it is not necessary to click the menu buttons each time. Instead we can write few SPSS commands called SPSS SYNTAX. For instance, a simple command to perform One-way ANOVA is

ANOVA SL25 BY TREATMENT(1,4), VARIETY(1,3)

This command has to be typed in SPSS with the commands File ▸ New ▸ Syntax. We may use more than one commands and each commands ends with '.' only. We can select the Run option available in the Syntax window.

12.4 OTHER USEFUL SOFTWARE

There are several other related software packages that are helpful to carry out data analysis. One such software is *S-Plus*. It has a large number of statistical routines including graphics. The only requirement is that it needs about 66 MB of disk space. It supports navigation across Window-based softwares.

Another software is *KyPLOT*. It needs only 18.4 MB of disk space and works in Windows environment. It supports several mathematical and statistical features. Working in it is much similar to working in Excel. The salient mathematical and statistical features of these softwares are listed as under:

Mathematical features	Statistical features
• Equation	• Descriptive statistics
• Integration	• Contingency tables
• Optimization	• Parametric tests
• Deconvolution	• Non-parametric tests
• Fourier transformation	• Regression analysis
• Spectral decomposition	• Multivariate analysis
• Time series analysis	
• Gabor transform	
• Wavelet analysis	
• Simulate random data	

In addition to these aspects, KyPLOT supports a large number of statistical graphs.

We close this chapter with the observation that advanced statistical analysis requires a full-fledged statistical software. Depending on the nature of work, we may use Excel in some cases and SPSS for advanced analysis.

REFERENCES

1. The Statistical Package for Social Sciences (SPSS), online help and Base System User's guide.
2. KyPLOT software, online system help.
3. S-Plus software, online help.

DO IT YOURSELF

12.1 Open SPSS. Try to open FoxPro file. See various types of files that can be opened by SPSS.

12.2 Once the data file is opened we can carry out data transformations, data filtering, etc. Consider any data file and work out on the options Edit ▸ Options ▸ Pivot Tables. You will find a number of styles for output.

12.3 Open the menu item Data and choose the option Select Cases. You will find an interesting option for filtering cases. Follow the instructions in the dialogue box. You will find that a crossed line will show the unselected data cases in the first column. We can use this tip to select only MALES from the file having both MALES and FEMALES. This is very similar to Excel filter.

12.4 Create a graph in SPSS and double click on it. You will find several options to modify its appearance. You can change the fonts, the filling style and so on.

12.5 A very important facility in SPSS is that of merging of data files. Suppose you have two separate files each having different fields with one or two common fields like SERIAL NUMBER. If both are SAV files, we can merge them using the options in DATA. This creates a combined file.

12.6 We can also save a SPSS file as a FoxPro file. This can be done by clicking the Save As option and choosing the file type as dBASE.

Appendices

APPENDIX A. DATA OF THE FILE C:\STATMAN\TRIBAL.DBF

SNO	SEX	CODE	AGE	HT	WT	BMI	BMR	PRO	FAT	CHO	CAL	IRON	VITA	VITC	EIN	EEX
1	1	1	21	158	51.0	20.43	1459.3	54.0	37.0	518.0	360	21	2321	31	2654.0	3674.2
2	1	1	21	152	52.0	22.51	1474.6	58.0	34.0	524.0	363	22	2481	32	2667.0	3697.9
3	1	1	22	159	48.0	18.99	1413.4	61.0	36.0	534.0	388	23	2436	28	2704.0	3373.4
4	1	1	23	157	50.5	20.49	1451.6	54.0	42.0	526.0	372	26	2151	29	2698.0	3579.3
5	1	1	24	157	57.0	23.12	1551.1	56.0	39.0	521.0	381	24	2122	27	2699.0	4066.1
6	1	1	25	153	60.0	25.63	1597.0	57.0	41.0	542.0	379	23	2238	31	2796.0	3992.7
7	1	1	25	156	44.0	18.08	1352.2	58.0	42.0	524.0	386	24	2408	36	2738.0	3059.9
8	1	1	25	158	51.5	20.63	1466.9	52.0	43.0	512.0	381	28	2326	35	2681.0	3605.6
9	1	1	26	159	57.0	22.55	1581.1	59.0	46.0	544.0	372	21	2129	28	2857.0	4002.9
10	1	1	27	158	56.0	22.43	1535.8	53.0	45.0	538.0	389	20	2189	27	2804.0	3890.1
11	1	1	27	154	54.0	22.77	1505.2	58.0	39.0	526.0	376	18	2326	29	2712.0	3799.3
12	1	1	28	158	58.0	23.23	1566.4	52.0	48.0	556.0	388	17	2289	26	2875.0	4067.2
13	1	1	29	161	59.0	22.76	1581.7	61.0	42.0	551.0	361	16	2186	32	2839.0	4060.8
14	1	1	29	159	57.5	22.74	1558.8	58.0	47.0	548.0	373	22	2381	34	2862.0	3947.2
15	1	1	30	157	49.5	20.08	1453.2	51.0	46.0	532.0	389	19	2286	33	2769.0	3346.0

APPENDIX A (contd.)

SNO	SEX	CODE	AGE	HT	WT	BMI	BMR	PRO	FAT	CHO	CAL	IRON	VITA	VITC	EIN	EEX
16	1	1	30	156	51.0	20.96	1470.6	42.0	41.0	489.0	348	21	2132	34	2492.0	3564.2
17	1	1	32	155	54.0	22.48	1505.4	49.0	40.0	512.0	347	23	2412	32	2624.0	3725.3
18	1	1	35	158	56.0	22.43	1528.6	51.0	42.0	551.0	386	26	2638	34	2798.0	3885.3
19	1	1	37	154	59.5	25.09	1569.2	59.0	46.0	552.0	346	25	2132	28	2879.0	4064.9
20	1	1	38	157	52.0	21.10	1482.2	57.0	44.0	542.0	343	22	2198	27	2816.0	3651.5
21	1	1	40	159	45.0	17.80	1401.0	49.0	38.0	498.0	361	21	2325	36	2548.0	3201.6
22	1	1	41	160	53.0	20.70	1493.8	51.0	42.0	512.0	379	29	2405	26	2649.0	3781.7
23	1	1	44	156	49.0	20.13	1447.4	49.0	44.0	511.0	386	21	2385	29	2652.0	3471.3
24	1	1	46	155	50.0	20.81	1459.0	46.0	41.0	508.0	363	18	2392	32	2585.0	3364.2
25	1	1	49	150	51.0	22.67	1470.6	45.0	39.0	499.0	399	16	2398	34	2549.0	3573.1
26	2	1	18	158	41.0	16.42	1098.7	55.1	28.1	353.2	318	28	1881	29	1918.1	2255.6
27	2	1	19	159	48.0	18.99	1201.6	51.9	30.2	354.7	371	26	1840	26	1924.0	2627.6
28	2	1	19	152	45.0	19.48	1157.5	52.8	25.2	352.2	361	22	1880	28	1879.5	2432.6
29	2	1	20	146	38.0	17.83	1054.6	42.0	20.0	346.0	301	20	1908	32	1759.8	2053.8
30	2	1	21	155	45.0	18.73	1157.5	62.3	26.4	361.5	311	20	1840	31	1967.4	2446.7
31	2	1	21	152	55.0	23.81	1304.5	42.3	29.4	437.0	318	21	1877	28	2196.4	2976.7
32	2	1	21	165	70.0	25.71	1525.0	57.1	20.3	537.5	316	20	1880	29	2608.0	3522.5
33	2	1	21	163	48.0	18.07	1201.6	55.0	28.9	350.0	323	19	1980	27	1916.6	2648.9
34	2	1	22	150	53.0	23.56	1275.1	64.7	23.9	401.2	341	21	1991	26	2121.9	2862.6
35	2	1	22	158	50.0	20.03	1231.0	62.3	26.9	375.0	333	22	2010	28	2028.3	2728.9
36	2	1	22	163	51.0	19.20	1245.7	54.2	40.2	336.2	341	26	2121	29	1963.4	2831.4
37	2	1	22	150	45.0	20.00	1157.5	62.2	34.2	356.2	352	23	2123	31	2005.5	2453.8
38	2	1	22	157	65.0	26.37	1451.5	54.2	32.0	421.6	351	22	2008	26	2220.9	3217.1
39	2	1	23	151	41.0	17.98	1098.7	64.4	22.8	374.6	363	21	1981	27	1984.0	2232.9
40	2	1	23	161	48.0	18.52	1201.6	47.4	23.6	366.2	361	19	1911	28	1884.6	2602.0

APPENDIX A (contd.)

SNO	SEX	CODE	AGE	HT	WT	BMI	BMR	PRO	FAT	CHO	CAL	IRON	VITA	VITC	EIN	EEX
41	2	1	24	145	44.0	20.93	1142.8	49.1	35.1	398.1	348	18	1963	29	2126.3	2376.3
42	2	1	25	145	41.0	19.50	1098.7	63.2	29.1	357.1	349	17	1989	18	1989.1	2249.9
43	2	1	26	153	49.0	20.93	1216.3	61.5	35.2	363.6	351	16	2120	27	2036.4	2669.6
44	2	1	30	163	55.0	20.70	1307.5	51.8	24.2	371.5	356	19	2230	21	2069.5	2929.4
45	2	1	31	145	44.0	20.93	1211.8	57.6	20.9	538.5	343	18	2180	22	2618.0	2398.4
46	2	1	33	156	50.0	20.55	1264.0	53.2	33.1	422.0	342	21	2190	23	2198.0	2715.1
47	2	1	33	154	45.0	18.97	1220.5	55.4	41.6	337.1	331	22	2240	27	1983.4	2479.2
48	2	1	34	142	49.0	24.30	1255.3	65.8	24.1	403.2	329	23	2260	22	2131.0	2659.0
49	2	1	35	156	45.0	18.49	1220.5	57.0	29.9	354.1	318	21	2315	28	1936.6	2529.8
50	2	1	35	146	43.0	20.17	1203.1	55.4	37.8	359.1	317	18	2415	29	2037.8	2335.1
51	1	2	22	154	42.0	17.70	1321.6	36.0	25.0	356.0	362	21	1860	29	1813.0	2614.0
52	1	2	22	153	50.0	21.40	1444.0	42.0	28.0	435.0	353	19	1865	28	2186.0	3382.8
53	1	2	23	154	49.0	20.70	1428.7	46.0	29.0	436.0	324	18	1932	27	2211.0	3345.7
54	1	2	25	151	44.0	19.30	1352.2	48.0	26.0	425.0	381	16	1981	24	2146.0	3025.1
55	1	2	26	159	53.0	21.00	1489.9	43.0	28.0	436.0	326	17	2021	26	2201.0	3583.2
56	1	2	26	153	42.0	17.90	1321.6	47.0	27.0	386.0	332	18	2091	29	1963.0	2927.0
57	1	2	26	152	43.0	18.60	1336.9	44.0	22.0	452.0	333	19	1986	31	2213.0	2994.7
58	1	2	26	167	51.5	18.50	1467.0	39.0	28.0	391.0	346	16	1884	34	1592.0	3468.6
59	1	2	28	159	46.0	18.20	1382.8	42.0	24.0	401.0	366	14	1784	24	2003.0	3156.8
60	1	2	28	169	58.0	20.30	1566.4	47.0	26.0	486.0	356	15	1942	26	2381.0	3869.8
61	1	2	30	154	49.0	20.70	1447.4	49.0	31.0	442.0	351	16	2041	29	2264.0	3250.5
62	1	2	31	163	54.0	20.30	1505.4	46.0	30.0	444.0	342	15	1842	32	2259.0	3610.6
63	1	2	32	162	51.5	19.60	1476.4	51.0	25.0	462.0	343	14	1921	33	2296.0	3519.0
64	1	2	32	159	48.0	19.00	1435.8	48.0	27.0	453.0	341	18	2122	32	2247.0	3348.7
65	1	2	32	159	49.0	19.40	1447.4	46.0	24.0	482.0	402	14	2132	31	2342.0	3270.3

APPENDIX A (contd.)

SNO	SEX	CODE	AGE	HT	WT	BMI	BMR	PRO	FAT	CHO	CAL	IRON	VITA	VITC	EIN	EEX
66	1	2	32	163	47.0	17.70	1424.2	42.0	26.0	481.0	363	19	2081	28	2351.0	3122.3
67	1	2	32	159	60.0	23.70	1575.0	46.0	27.0	462.0	381	21	1981	27	2294.0	4008.1
68	1	2	35	162	50.0	19.10	1459.0	47.0	28.0	471.0	342	17	1786	26	2212.0	3385.8
69	1	2	36	163	51.0	19.20	1470.6	49.0	32.0	498.0	351	14	1742	28	2491.0	3431.8
70	1	2	36	159	50.0	19.80	1459.0	48.0	29.0	481.0	361	15	1692	29	2401.0	3380.3
71	1	2	46	161	49.0	18.90	1447.4	46.0	33.0	479.0	363	16	1821	25	2412.0	3327.5
72	1	2	49	163	54.0	20.30	1505.4	49.0	30.0	462.0	354	17	1724	24	2341.0	3668.3
73	1	2	50	159	48.0	19.00	1435.8	43.0	27.0	456.0	356	19	1732	23	2236.0	3264.2
74	1	2	52	160	48.0	18.80	1435.8	43.0	29.0	422.0	358	20	1745	21	2142.0	3297.8
75	1	2	55	162	47.0	17.90	1424.2	47.0	28.0	404.0	354	21	1892	29	2072.0	3269.9
76	2	2	20	151	40.5	17.80	1091.4	34.0	15.4	306.0	281	14	1818	17	1504.0	2226.2
77	2	2	20	149	32.5	14.60	973.8	32.8	15.2	301.1	289	15	1820	19	1486.6	1776.6
78	2	2	20	148	43.0	19.60	1128.1	38.2	15.0	320.1	291	17	1818	18	1585.3	2300.4
79	2	2	20	149	35.5	16.00	1017.9	40.0	20.0	287.6	298	18	1836	17	1506.8	2090.4
80	2	2	22	150	41.5	18.40	1106.1	35.0	16.5	300.0	297	16	1810	16	1540.0	2262.2
81	2	2	23	151	40.5	17.80	1091.4	41.0	23.0	310.0	298	14	1910	19	1590.0	2279.3
82	2	2	23	150	42.0	18.70	1113.4	39.0	18.5	310.0	299	13	1780	20	1580.0	2361.1
83	2	2	24	148	39.0	17.80	1069.3	40.5	20.5	305.5	296	12	1796	18	1600.0	2199.3
84	2	2	24	147	40.5	18.70	1091.4	34.0	15.0	324.0	289	13	1792	19	1588.0	2286.9
85	2	2	25	151	44.0	19.30	1142.8	46.5	28.0	321.0	286	14	1798	20	1700.0	2476.3
86	2	2	25	152	39.0	16.90	1069.3	36.5	12.0	332.6	289	15	1881	19	1586.6	2296.5
87	2	2	25	153	47.0	20.10	1186.9	39.5	13.0	305.0	293	16	1821	18	1940.5	2600.7
88	2	2	27	156	43.0	17.70	1128.1	33.7	14.2	333.2	292	14	1811	17	1664.7	2476.8
89	2	2	29	155	49.0	20.40	1216.3	44.0	26.2	311.2	293	13	1908	18	1561.0	2964.5

APPENDIX A (contd.)

SNO	SEX	CODE	AGE	HT	WT	BMI	BMR	PRO	FAT	CHO	CAL	IRON	VITA	VITC	EIN	EEX
90	2	2	29	157	50.0	20.30	1231.0	39.0	13.0	306.0	294	14	1816	19	1940.0	2730.3
91	2	2	30	154	49.0	20.70	1255.3	37.0	21.0	348.0	296	12	1814	19	1904.0	2720.4
92	2	2	30	153	45.0	19.20	1220.5	61.0	33.0	380.0	295	13	1811	18	2021.0	2512.2
93	2	2	30	156	48.0	19.70	1246.6	43.0	26.0	406.0	301	12	1806	17	2024.0	2669.4
94	2	2	31	153	39.0	16.70	1168.3	31.0	16.0	240.0	304	13	1809	18	1599.0	2300.2
95	2	2	35	156	50.0	20.50	1264.0	40.8	40.5	394.0	306	14	1804	18	2088.0	2676.7
96	2	2	36	157	42.0	17.00	1194.4	51.0	43.0	400.0	305	15	1801	19	2006.0	2414.2
97	2	2	36	151	38.0	16.70	1159.6	34.0	12.0	238.0	304	16	1803	16	1489.6	2196.6
98	2	2	38	152	43.0	18.60	1203.1	60.5	32.0	400.0	303	17	1806	18	2085.0	2354.9
99	2	2	42	153	46.0	19.70	1229.2	59.0	23.0	320.0	306	19	1803	19	2030.0	2552.0
100	2	2	42	156	44.0	18.10	1211.8	62.0	25.0	351.0	309	18	1811	18	2040.0	2526.3
101	1	3	25	152	58.0	25.10	1566.4	49.0	38.0	498.0	382	21	2051	29	2548.0	4223.4
102	1	3	27	156	53.0	21.80	1489.9	47.0	26.0	486.0	379	23	2122	21	2381.0	3689.6
103	1	3	28	159	50.0	19.80	1444.0	42.0	26.0	481.0	361	20	2238	26	2351.0	3591.1
104	1	3	28	160	47.0	18.40	1398.1	41.0	25.0	376.0	374	16	2029	27	2013.0	3318.5
105	1	3	29	158	47.0	18.80	1398.1	46.0	29.0	436.0	365	21	2189	29	2211.0	3376.2
106	1	3	29	157	51.0	20.70	1459.3	49.0	32.0	498.0	366	22	2289	24	2491.0	3720.7
107	1	3	29	158	47.0	18.80	1398.1	47.0	28.0	471.0	363	17	2186	29	2212.0	3418.5
108	1	3	31	159	49.0	19.40	1447.4	48.0	29.0	481.0	361	23	2032	28	2401.0	3432.4
109	1	3	31	161	48.0	18.50	1435.8	43.0	27.0	456.0	372	19	2198	29	2236.0	3459.3
110	1	3	33	160	47.0	18.40	1424.2	45.0	28.0	406.0	379	17	1932	31	2072.0	3356.0
111	1	3	33	156	46.0	18.90	1412.6	47.0	30.0	464.0	383	21	2061	32	2341.0	3377.2
112	1	3	34	157	47.0	19.10	1424.2	45.0	24.0	483.0	365	22	2021	33	2342.0	3353.1
113	1	3	34	154	49.0	20.70	1447.4	45.0	38.0	497.0	346	19	1981	31	2547.0	3484.5
114	1	3	35	156	51.0	21.00	1470.6	45.0	42.0	507.0	374	22	2036	32	2579.0	3577.9
115	1	3	36	163	52.0	19.60	1482.2	46.0	36.0	492.0	359	23	2143	34	2512.0	3710.8
116	1	3	36	169	57.0	20.00	1540.2	52.0	31.0	452.0	369	21	2324	28	2319.0	4118.9

APPENDIX A (contd.)

SNO	SEX	CODE	AGE	HT	WT	BMI	BMR	PRO	FAT	CHO	CAL	IRON	VITA	VITC	EIN	EEX
117	1	3	38	163	58.0	21.80	1551.8	51.0	29.0	462.0	358	23	2021	27	2336.0	4153.4
118	1	3	38	159	50.0	19.80	1459.0	47.0	32.0	479.0	354	23	2091	31	2412.0	3540.8
119	1	3	38	155	50.0	20.80	1459.0	52.0	31.0	496.0	364	21	2041	32	2487.0	3618.4
120	1	3	39	158	50.0	20.00	1459.0	49.0	31.0	484.0	353	22	2022	34	2429.0	3592.1
121	1	3	39	165	50.0	18.40	1459.0	52.0	34.0	476.0	346	24	2132	29	2435.0	3572.7
122	1	3	40	166	53.0	19.20	1493.8	46.0	24.0	461.0	342	19	1821	28	2269.0	3825.7
123	1	3	41	162	53.0	20.20	1493.8	47.0	31.0	489.0	351	26	1892	27	2446.0	3778.0
124	1	3	42	169	54.0	18.90	1505.4	49.0	29.0	462.0	368	23	1986	33	2321.0	3924.8
125	1	3	43	158	51.0	20.40	1470.6	49.0	32.0	501.0	359	27	2081	36	2502.0	3732.4
126	2	3	21	152	45.0	19.50	1157.5	46.9	27.3	327.6	301	19	1880	21	1771.8	2473.6
127	2	3	21	151	35.0	15.40	1010.5	44.4	24.2	301.2	311	17	1860	22	1687.4	1906.6
128	2	3	22	149	35.0	15.80	1010.5	44.8	30.3	297.5	314	18	1870	24	1662.3	1872.4
129	2	3	24	150	39.0	17.30	1069.3	54.3	32.8	314.5	306	16	1830	26	1789.5	2099.8
130	2	3	24	149	43.0	19.40	1128.1	46.9	24.8	380.0	304	19	1860	27	1984.2	2379.5
131	2	3	26	151	48.0	21.10	1201.6	64.0	22.0	374.0	306	20	1863	28	1984.0	2711.4
132	2	3	26	148	43.0	19.60	1128.1	54.0	40.0	336.0	309	19	1871	29	1963.0	2327.3
133	2	3	26	150	38.0	16.90	1054.6	44.3	22.2	341.6	311	18	1872	26	1766.1	2006.1
134	2	3	26	153	45.0	19.20	1157.5	36.2	16.9	386.9	314	17	1873	25	1862.1	2510.9
135	2	3	27	151	51.0	22.40	1245.7	55.0	33.0	422.0	316	18	1893	24	2239.0	2937.7
136	2	3	27	149	40.0	18.00	1084.0	54.0	33.0	316.0	318	19	1889	23	1798.6	2184.8
137	2	3	27	153	44.0	18.80	1142.8	42.0	26.0	340.0	319	20	1899	22	1781.0	2503.1
138	2	3	28	154	48.0	20.20	1201.6	40.8	31.0	311.6	321	21	1898	21	1790.9	2826.4
139	2	3	29	152	46.0	19.90	1172.2	46.8	24.9	344.1	331	19	1992	23	1789.0	2567.9

APPENDIX A (contd.)

SNO	SEX	CODE	AGE	HT	WT	BMI	BMR	PRO	FAT	CHO	CAL	IRON	VITA	VITC	EIN	EEX
140	2	3	29	151	49.0	21.50	1216.3	41.0	19.8	345.4	333	18	1962	26	1734.0	2797.4
141	2	3	30	153	48.0	20.50	1246.6	54.6	29.9	317.0	346	17	1921	25	1754.0	2710.9
142	2	3	31	149	40.0	18.00	1177.0	47.1	26.8	326.0	341	16	1931	24	1774.6	2139.4
143	2	3	32	148	49.0	22.40	1255.3	43.0	29.0	437.0	321	17	1939	23	2196.0	2763.3
144	2	3	35	149	35.0	15.80	1133.5	43.0	22.0	307.0	326	19	1938	22	1636.0	1969.7
145	2	3	35	148	56.0	25.60	1316.2	57.0	20.0	537.0	321	18	1912	19	2608.0	2948.3
146	2	3	37	151	47.0	20.60	1237.9	59.0	25.8	423.0	333	19	1892	18	2321.0	2843.5
147	2	3	38	151	48.0	21.10	1246.6	62.0	26.0	362.0	356	18	1862	21	1969.0	2614.5
148	2	3	39	153	51.0	21.80	1272.7	50.0	33.0	386.0	309	19	1932	22	2056.0	2785.0
149	2	3	39	152	50.0	21.60	1264.0	57.0	23.0	398.0	308	17	1912	21	2034.0	2808.6
150	2	3	40	149	44.0	19.80	1211.8	53.0	26.0	371.0	311	18	1908	20	2089.0	2405.3

APPENDIX B. DATA OF THE FILE C:\STATMAN\BLOOD.DBF

SNO	YEAR	MONTH	OPOS	ONEG	APOS	ANEG	BPOS	BNEG	ABPOS	ABNEG	RHPOS	RHNEG	TOT
1	1993	MAR	8	1	4	0	6	1	0	0	18	2	20
2	1993	APR	14	2	0	0	8	0	0	1	22	3	25
3	1993	MAY	28	2	6	4	4	0	1	0	39	6	45
4	1993	JUN	35	0	4	0	7	0	1	0	47	0	47
5	1993	JUL	30	0	19	0	26	3	5	0	80	3	83
6	1993	AUG	4	2	1	2	9	0	0	0	14	4	18
7	1993	SEP	42	0	20	0	14	0	5	1	81	1	82
8	1993	OCT	58	1	29	0	67	3	5	0	159	4	163
9	1993	NOV	69	19	41	0	39	5	8	1	157	25	182
10	1993	DEC	93	3	11	0	47	7	6	1	157	11	168
11	1994	JAN	60	0	39	4	46	1	13	4	158	9	167
12	1994	FEB	94	6	20	0	71	6	4	0	189	12	201
13	1994	MAR	131	18	40	4	79	5	15	3	265	30	295
14	1994	APR	97	12	49	6	54	5	16	4	216	27	243
15	1994	MAY	119	3	39	3	75	0	13	2	246	8	254
16	1994	JUN	161	5	45	6	114	10	11	1	331	22	353
17	1994	JUL	113	14	43	7	75	16	18	1	249	38	287
18	1994	AUG	113	12	50	5	80	4	20	3	263	24	287
19	1994	SEP	109	6	41	3	69	9	13	0	232	18	250
20	1994	OCT	85	6	34	3	56	2	11	0	186	11	197
21	1994	NOV	75	4	37	4	73	6	12	3	197	17	214
22	1994	DEC	78	3	33	5	48	6	16	1	175	15	190
23	1995	JAN	131	8	53	4	93	5	10	1	287	18	305
24	1995	FEB	94	7	54	1	51	8	19	3	218	19	237
25	1995	MAR	102	13	45	1	76	6	12	1	235	21	256

APPENDIX B (contd.)

26	1995	APR	82	4	44	1	54	3	13	0	193	8	201
27	1995	MAY	98	5	33	6	59	6	5	11	195	28	223
28	1995	JUN	83	2	35	3	40	4	9	2	167	11	178
29	1995	JUL	73	2	21	3	31	8	20	0	145	13	158
30	1995	AUG	96	12	40	9	59	9	20	2	215	32	247
31	1995	SEP	117	11	46	9	68	13	15	0	246	33	279
32	1995	OCT	95	13	55	6	72	7	12	0	234	26	260
33	1995	NOV	75	10	37	4	75	4	12	0	199	18	217
34	1995	DEC	89	7	39	1	54	9	7	1	189	18	207
35	1996	JAN	68	7	42	4	43	7	12	1	165	19	184
36	1996	FEB	130	14	40	5	78	6	20	3	268	28	296
37	1996	MAR	100	11	43	2	65	5	18	2	226	20	246
38	1996	APR	84	8	28	3	76	11	18	2	206	24	230
39	1996	MAY	133	21	57	5	94	3	9	3	293	32	325
40	1996	JUN	58	5	22	3	47	2	7	3	134	13	147
41	1996	JUL	94	10	56	8	62	4	12	2	224	24	248
42	1996	AUG	58	3	20	9	44	3	10	1	132	16	148
43	1996	SEP	91	9	35	2	60	7	8	0	194	18	212
44	1996	OCT	87	7	64	1	71	5	10	0	232	13	245
45	1996	NOV	80	4	44	5	63	7	10	2	197	18	215
46	1996	DEC	47	6	26	2	50	3	4	1	127	12	139

APPENDIX C. DATA OF THE FILE C:\STATMAN\PLANT.DBF

SNO	VARIETY	TREATMENT	SL-25	SL-35	SL-45	SL-55
1	1	1	15	20	29	40
2	1	1	16	21	30	41
3	1	1	15	21	29	40
4	1	1	14	22	31	42
5	1	1	15	21	30	41
6	1	2	19	28	40	50
7	1	2	17	27	41	51
8	1	2	16	29	42	52
9	1	2	17	28	41	50
10	1	2	19	27	40	51
11	1	3	20	36	46	56
12	1	3	21	35	45	57
13	1	3	19	34	47	58
14	1	3	20	35	46	56
15	1	3	21	36	45	59
16	1	4	15	44	54	60
17	1	4	24	43	55	59
18	1	4	23	44	56	61
19	1	4	25	45	55	62
20	1	4	24	43	56	61
21	2	1	17	25	38	48
22	2	1	16	24	35	46
23	2	1	17	26	36	47
24	2	1	16	24	37	49
25	2	1	17	25	39	48
26	2	2	20	30	43	54
27	2	2	21	32	44	55
28	2	2	19	31	45	56
29	2	2	20	33	46	57
30	2	2	21	32	45	58
31	2	3	23	40	50	59
32	2	3	22	41	51	60
33	2	3	24	42	52	61
34	2	3	23	41	53	60
35	2	3	24	42	50	62
36	2	4	27	51	59	63
37	2	4	28	52	60	64
38	2	4	29	54	61	65
39	2	4	30	55	59	66
40	2	4	32	56	58	67
41	3	1	15	20	28	46
42	3	1	16	21	29	47
43	3	1	14	20	30	48
44	3	1	15	21	31	45
45	3	1	14	22	32	44
46	3	2	18	29	40	54

APPENDIX C (contd.)

SNO	VARIETY	TREATMENT	SL25	SL35	SL45	SL55
47	3	2	17	30	41	53
48	3	2	16	29	42	55
49	3	2	17	31	41	56
50	3	2	18	29	40	54
51	3	3	20	38	48	58
52	3	3	21	37	49	57
53	3	3	19	36	47	56
54	3	3	21	39	43	55
55	3	3	22	38	45	57
56	3	4	26	46	54	64
57	3	4	27	45	55	63
58	3	4	28	44	56	62
59	3	4	24	45	57	61
60	3	4	23	46	54	63

APPENDIX D. VALUES OF STUDENT'S t-DISTRIBUTION FOR DIFFERENT DEGREES OF FREEDOM (df) AT SELECTED LEVEL OF SIGNIFICANCE (α)

(Prepared with the help of the Excel paste function TINV ( ))

df	$\alpha = 0.100$	$\alpha = 0.050$	$\alpha = 0.025$	$\alpha = 0.010$	$\alpha = 0.005$
1	3.0777	6.3137	12.7062	31.8210	63.6559
2	1.8856	2.9200	4.3027	6.9645	9.9250
3	1.6377	2.3534	3.1824	4.5407	5.8408
4	1.5332	2.1318	2.7765	3.7469	4.6041
5	1.4759	2.0150	2.5706	3.3649	4.0321
6	1.4398	1.9432	2.4469	3.1427	3.7074
7	1.4149	1.8946	2.3646	2.9979	3.4995
8	1.3968	1.8595	2.3060	2.8965	3.3554
9	1.3830	1.8331	2.2622	2.8214	3.2498
10	1.3712	1.8125	2.2281	2.7638	3.1693
11	1.3634	1.7959	2.2010	2.7181	3.1058
12	1.3562	1.7823	2.1788	2.6810	3.0545
13	1.3502	1.7709	2.1604	2.6503	3.0123
14	1.3450	1.7613	2.1448	2.6245	2.9768
15	1.3406	1.7531	2.1315	2.6025	2.9467
16	1.3368	1.7459	2.1199	2.5835	2.9208
17	1.3334	1.7396	2.1098	2.5669	2.8982
18	1.3304	1.7341	2.1009	2.5524	2.8784
19	1.3277	1.7291	2.0930	2.5395	2.8609
20	1.3253	1.7247	2.0860	2.5280	2.8453
21	1.3232	1.7207	2.0796	2.5176	2.8314
22	1.3212	1.7171	2.0739	2.5083	2.8188
23	1.3195	1.7139	2.0687	2.4999	2.8073
24	1.3178	1.7109	2.0639	2.4922	2.7970
25	1.3163	1.7081	2.0595	2.4851	2.7874
26	1.3150	1.7056	2.0555	2.4786	2.7787
27	1.3137	1.7033	2.0518	2.4727	2.7707
28	1.3125	1.7011	2.0484	2.4671	2.7633
29	1.3114	1.6991	2.0452	2.4620	2.7564
30	1.3104	1.6973	2.0423	2.4573	2.7500
∞	1.3095	1.6955	2.0395	2.4528	2.7440

APPENDIX E. CUMULATIVE STANDARD NORMAL DISTRIBUTION VALUES FROM $-\infty$ TO z

(Prepared with the help of the paste function NORMSDIST ( ) for the different values of z)

z	0.00	0.01	0.02	0.03	0.04	0.05	0.06	0.07	0.08	0.09
0.00	0.5000	0.5040	0.5080	0.5120	0.5160	0.5199	0.5239	0.5279	0.5319	0.5359
0.10	0.5398	0.5438	0.5478	0.5517	0.5557	0.5596	0.5636	0.5675	0.5714	0.5753
0.20	0.5793	0.5832	0.5871	0.5910	0.5948	0.5987	0.6026	0.6064	0.6103	0.6141
0.30	0.6179	0.6217	0.6255	0.6293	0.6331	0.6368	0.6406	0.6443	0.6480	0.6517
0.40	0.6554	0.6591	0.6628	0.6664	0.6700	0.6736	0.6772	0.6808	0.6844	0.6879
0.50	0.6915	0.6950	0.6985	0.7019	0.7054	0.7088	0.7123	0.7157	0.7190	0.7224
0.60	0.7257	0.7291	0.7324	0.7357	0.7389	0.7422	0.7454	0.7486	0.7517	0.7549
0.70	0.7580	0.7611	0.7642	0.7673	0.7704	0.7734	0.7764	0.7794	0.7823	0.7852
0.80	0.7881	0.7910	0.7939	0.7967	0.7995	0.8023	0.8051	0.8078	0.8106	0.8133
0.90	0.8159	0.8186	0.8212	0.8238	0.8264	0.8289	0.8315	0.8340	0.8365	0.8389
1.00	0.8413	0.8438	0.8461	0.8485	0.8508	0.8531	0.8554	0.8577	0.8599	0.8621
1.10	0.8643	0.8665	0.8686	0.8708	0.8729	0.8749	0.8770	0.8790	0.8810	0.8830
1.20	0.8849	0.8869	0.8888	0.8907	0.8925	0.8944	0.8962	0.8980	0.8997	0.9015
1.30	0.9032	0.9049	0.9066	0.9082	0.9099	0.9115	0.9131	0.9147	0.9162	0.9177
1.40	0.9192	0.9207	0.9222	0.9236	0.9251	0.9265	0.9279	0.9292	0.9306	0.9319
1.50	0.9332	0.9345	0.9357	0.9370	0.9382	0.9394	0.9406	0.9418	0.9429	0.9441
1.60	0.9452	0.9463	0.9474	0.9484	0.9495	0.9505	0.9515	0.9525	0.9535	0.9545
1.70	0.9554	0.9564	0.9573	0.9582	0.9591	0.9599	0.9608	0.9616	0.9625	0.9633
1.80	0.9641	0.9649	0.9656	0.9664	0.9671	0.9678	0.9686	0.9693	0.9699	0.9706
1.90	0.9713	0.9719	0.9726	0.9732	0.9738	0.9744	0.9750	0.9756	0.9761	0.9767
2.00	0.9772	0.9778	0.9783	0.9788	0.9793	0.9798	0.9803	0.9808	0.9812	0.9817

Appendix E (Contd.)

z	0.00	0.01	0.02	0.03	0.04	0.05	0.06	0.07	0.08	0.09
2.10	0.9821	0.9826	0.9830	0.9834	0.9838	0.9842	0.9846	0.9850	0.9854	0.9857
2.20	0.9861	0.9864	0.9868	0.9871	0.9875	0.9878	0.9881	0.9884	0.9887	0.9890
2.30	0.9893	0.9896	0.9898	0.9901	0.9904	0.9906	0.9909	0.9911	0.9913	0.9916
2.40	0.9918	0.9920	0.9922	0.9925	0.9927	0.9929	0.9931	0.9932	0.9934	0.9936
2.50	0.9938	0.9940	0.9941	0.9943	0.9945	0.9946	0.9948	0.9949	0.9951	0.9952
2.60	0.9953	0.9955	0.9956	0.9957	0.9959	0.9960	0.9961	0.9962	0.9963	0.9964
2.70	0.9965	0.9966	0.9967	0.9968	0.9969	0.9970	0.9971	0.9972	0.9973	0.9974
2.80	0.9974	0.9975	0.9976	0.9977	0.9977	0.9978	0.9979	0.9979	0.9980	0.9981
2.90	0.9981	0.9982	0.9982	0.9983	0.9984	0.9984	0.9985	0.9985	0.9986	0.9986
3.00	0.9987	0.9987	0.9987	0.9988	0.9988	0.9989	0.9989	0.9989	0.9990	0.9990
3.10	0.9990	0.9991	0.9991	0.9991	0.9992	0.9992	0.9992	0.9992	0.9993	0.9993
3.20	0.9993	0.9993	0.9994	0.9994	0.9994	0.9994	0.9994	0.9995	0.9995	0.9995
3.30	0.9995	0.9995	0.9995	0.9996	0.9996	0.9996	0.9996	0.9996	0.9996	0.9997
3.40	0.9997	0.9997	0.9997	0.9997	0.9997	0.9997	0.9997	0.9997	0.9997	0.9998
3.50	0.9998	0.9998	0.9998	0.9998	0.9998	0.9998	0.9998	0.9998	0.9998	0.9998
3.60	0.9998	0.9998	0.9999	0.9999	0.9999	0.9999	0.9999	0.9999	0.9999	0.9999
3.70	0.9999	0.9999	0.9999	0.9999	0.9999	0.9999	0.9999	0.9999	0.9999	0.9999
3.80	0.9999	0.9999	0.9999	0.9999	0.9999	0.9999	0.9999	0.9999	0.9999	0.9999
3.90	1.0000	1.0000	1.0000	1.0000	1.0000	1.0000	1.0000	1.0000	1.0000	1.0000
4.00	1.0000	1.0000	1.0000	1.0000	1.0000	1.0000	1.0000	1.0000	1.0000	1.0000

Index

Access, 32
Alt + Enter, 65
Analysis of variance (ANOVA), 145, 235
APPEND, 46, 49
Attributes, 13
Auto
  format, 78
  text, 77, 79
  width, 78
Autofit selection, 87
AVG( ), 52
Axes, 105

Bar chart, 102
Beta coefficients, 176
Bin, 118
Bivariate regression model, 166
Booting, 36
Browse, 38
Bullets, 83
Bytes, 33

Causal relationship, 166
CD-ROM, 34
Chart area, 103
CHI, 201, 203
Chi-square test, 154, 195
CLOSE ALL, 44
CNT( ), 52
Coefficient
  of determination, 163
  of variation, 21
Confidence interval, 118, 131
Contingency table, 14, 157
CORR, 199

Correlation
  analysis, 162
  coefficient, 162
  matrix, 166
Critical value, 133
Cross tabulation, 14, 122, 193
CROSSTABS, 193
Ctrl + A, 73
Ctrl + W, 44, 183

Damping factor, 216
Data
  Analysis Pak, 114
  labels, 104, 106
dBASEIII, IV, 44
Degrees of freedom (DF), 135
DELE, 48
Descriptive statistics, 115
Desktop, 66
DIR, 44
DISPLAY STRUCTURE, 47
Distribution, 224
DO, 183
DSTAT, 184

EDIT, 48
Equation editor, 81
Estimate, 130
Estimator, 130, 131
Excel, 31
Experimental method, 7
Exponential model, 206, 213

File, 30
  properties, 72
  save, 82

INDEX

Filter, 94
Financial functions, 221
Find, 70
Folder, 68
Fonts, 78
Forecast, 208
Format cells, 87
FoxBASE, 44
FoxPro, 38
Freeze pans, 90
Frequency distribution, 15, 118
FTAB, 189
F-test, 144
FV, 221

Gigabyte, 33
GO BOTT, 58
Goodness of fit, 154
GO TOP, 50
Graphic User Interface (GUI), 65
Gridlines, 103

Hard disk, 33
Hardware, 30
Help, 67
Histogram, 15, 118
Homoscedastic, 145
Hypothesis, 131, 132

Icons, 66
INSERT, 49
Interaction, 146, 150
Internet explorer, 68

Keyboard, 32
Kilobyte, 33
Kurtosis, 21, 24

Landscape, 111
Legend, 102, 104
Level of significance (LOS), 132
Line chart, 107
LIST, 50
Location of a graph, 103

Mahalanobis, 2

Mail merge, 79
Matrix inverse, transpose, 96
MAX( ), 52
Mean, 24
Median, 19, 24
Megabyte, 33
MIN( ), 52
Mode, 19, 24
Modem, 35
MODI COMM, 183
MODI STRUCTURE, 48
Mouse, 32
MS-Access, 75
MS-Excel, 18, 75
MS-Office, 31, 65, 74
MS-PowerPoint, 75
MS-Word, 75
Multicollinearity, 182
Multimedia, 35
Multiple linear regression, 173
My Computer, 68

Navigate, 68
Nested classification, 15
Normal
  curve, 16
  test, 135
NORMDIST, 226

Office toolbars, 75
One-factor ANOVA, 145
One-sample t-test, 136
Operating system, 30
Operations research, 28

PACK, 49
Paired t-test, 142
Parameters, 23
Pareto diagram, 119
Paste functions, 92
Pattern, 105
Personal computer, 29
Pie
  chart, 106, 107
  diagram, 16
Pivot tables, 122
Plot area, 103
Polynomial model, 206
Population, 5, 23

Portrait, 111
Power
  curve, 212
  model, 206
  of a test, 132
PowerPoint, 32
PRG files, 183
Printer, 34
Probability, 126
  distribution, 127
Purposive sample, 5
p-value, 134

Quit, 44

R^2-value, 167
Random
  numbers, 224
  number seed, 224
  sample, 5
  variable, 126
Range, 15, 20
Ranking, 222
RECALL, 49
RECCOUNT ( ), 53
Record, 42
Recycle bin, 66, 68, 73
Regression analysis, 162
Relative frequency, 126
REPLACE, 49
REPL ALL, 56
Replication, 146
Residuals, 177
Ruler, 77
RUN, 188

Sampling distribution, 25
Sankhya, 2
Scatter diagram, 101, 105, 162
SET TALK ON, 49
Skewness, 21, 24
Smoothing factor, 216
Software, 30
SORT, 49
SPSS, 229
Standard
  deviation, 20
  error, 25, 62, 115, 117, 130

Statistics, 130
Statistical functions, 93
STATMAN, 42, 54
STD( ), 52
Stratified random sample, 7
Student's t-distribution, 135
Sturge's formula, 16, 189
Subscripts, 80
SUM( ), 52
Summary statistics, 19, 115, 230
Survey method, 4
Symbols, 80

Tables, 78
Tabulation, 9
Task bar, 66
Templates, 75
Trend analysis, 205
TRIBAL, 44
TTEST, 141, 196
Two
  -factor ANOVA, 146, 151
  -sample t-test, 140
  -sample Z-test, 138
Type-1 error, 132
Type-2 error, 132

USE, 45

VAL( ), 51
VAR( ), 52
Variance, 20

WIN, 65
Windows explorer, 68
WordStar, 31
Workbook, 85
Worksheet, 75
  functions, 92

Y-bar errors, 106

Z-test, 135